D1224114

THE GREAT APES

THE GREAT APES

DAN FREEMAN

G P PUTNAM'S SONS
NEW YORK
A BISON BOOK

Published by
G. P. Putnam's Sons
200 Madison Avenue
New York, NY 10016
USA

Produced by
Bison Books Limited
4 Cromwell Place
London SW7

Library of Congress Catalog Card Number 79-84552
ISBN 399-12399-7

Printed in Hong Kong

Editor: John Man
Designer: Laurence Bradbury

CONTENTS

1/OF APES AND MEN

In evolutionary terms, *Homo sapiens* is one child in the family of apes, along with gorillas, orang-utans and chimpanzees – the Great Apes proper – and the Lesser Apes, the gibbons and siamangs. Researchers like Adrien Deschryver, seen at the left with wild gorillas in Zaire, have shown that we have much in common with the great apes. But how close is the relationship? To answer the question – though the details are still controversial – involves a look at the apes' family tree and at the intelligence of the most social of the non-human apes, the chimpanzee.

As man's closest living relatives in the animal world, the apes have taken on a special role in our lives. Their 'human-ness' has often been overexaggerated, perhaps in sentimental attempts to close the evolutionary gap between them and ourselves. At other times the differences have been emphasized to accentuate our own uniqueness.

Let us first of all be certain of our evolutionary ground. There are far more physical similarities linking apes and man than there are differences dividing them. The differences are represented by no more than some slight anatomical changes and a more advanced brain (from which of course, flow all the complexities of our social and cultural behavior).

In the simplest terms, we are, as a species, about as closely related to the gorilla and chimpanzee (the orang-utan is a rather more distant relative) as it is possible to be. We are, as it were, a mere generation removed from a common ancestor, when compared with the full evolutionary tree of life on earth. Traces of life have been detected in rocks 3000 million years old – over half the age of the Earth. One tenth of that span of time – 300 million years – takes us back to the age of amphibians, before the evolution of the dinosaurs. One tenth of that – 30 million years – and we are in the age of mammals. By then primates – the group to which both apes and men belong – had evolved, but we have to come forward to a final tenth of our original span – three million years – to bring us into the time of the first true humans.

The common ancestry of man and ape, therefore, is quite extensive. Among primates – that group of mammals containing the tarsiers, lorises, bushbabies, lemurs, monkeys, apes and man – the problems of working out an evolutionary history has been made very much easier by the continued existence of a whole range of animals sufficiently related to show the 'steps' by which the higher primates have evolved from an essentially primitive stock. Of course many of the 'steps' have become extinct over the years but those that still exist indicate the way the group must have evolved.

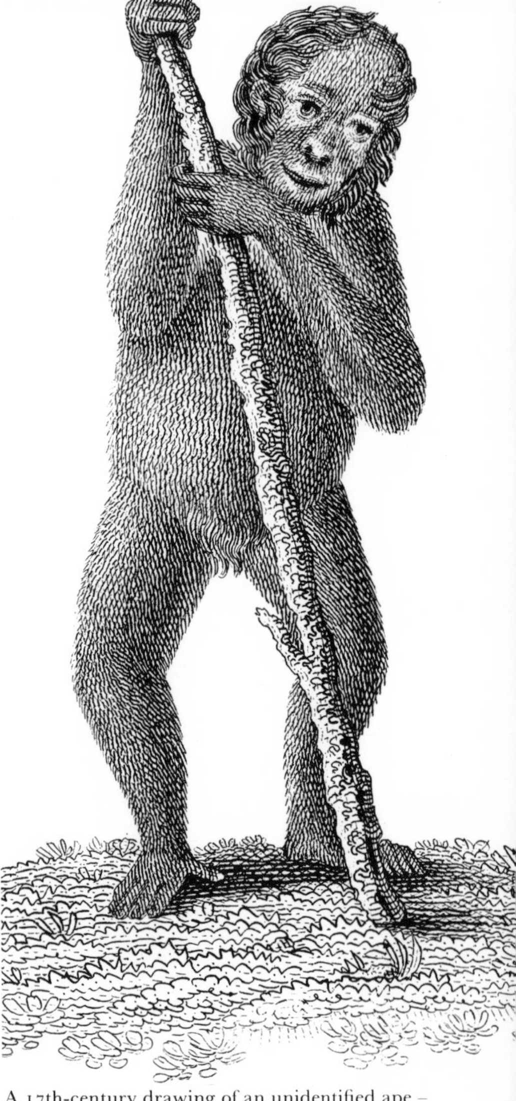

A 17th-century drawing of an unidentified ape – probably a chimpanzee – as a subhuman reflects the centuries-old recognition of the apes' family resemblance to man.

Even the earliest animals – the tree shrews – from which all primates may have evolved are still to be found scurrying around in the dense forests of Southeast Asia. A typical tree shrew is no larger than a squirrel and species have been in existence, largely unchanged, since at least the late Cretaceous – that is for about 70 million years. At that distant time there were no monkeys or apes – indeed hardly any birds or mammals at all. The dominant creatures of the day were reptiles, in particular the dinosaurs. As the dinosaurs crashed into extinction after an overwhelmingly successful reign spanning nearly 150 million years, birds and mammals began to flourish. Among them was our small, insect-eating ancestral primate.

It seems possible that the emerging mammals – particularly the rodents – were in fierce competition for both food and space. Some of them would have begun to climb into the lower branches of bushes to find safety and additional food. Hidden from ground-level competition, these small mammals would have bred more successfully than before and passed their tree-climbing characteristics on to their own young. The resulting spread to colonize a new habitat must have been very rapid. It was also very significant, for it gave rise to a whole new line of evolution and paved the way for the future emergence of many species of the primate that are found in the world today.

Because they show such a mixture of primate and non-primate characteristics, the classification of tree shrews remains controversial. Here are a few non-primate characteristics, with contrary arguments in parentheses:

– they do not look at all like primates (but related species do not neces-

sarily resemble each other closely);
- they do not rely primarily on vision (but their brain shows a significant enlargement in the area controlling vision);
- they do not show opposability – the free use of thumb and big toe (but they show a definite trend toward this);
- they have claws instead of nails on their digits (but one might expect that of representatives of such an early stage of primate evolution);
- and the females have multiple pairs of breasts instead of the usual one pair (but these are more reduced than in other primitive mammals).

There are many other reasons both for and against, but these few serve to show how thorny this particular problem is. No wonder some biologists think it safer to omit tree shrews from the primate family tree. The furthest we can safely go is to say that the ancestral primate species must have been rather similar to today's tree shrews in its looks but perhaps different in its behavior.

Our ancestral tree shrew-like species, then, moved off the ground and into the trees and flourished. To travel further afield and as safely as possible, some of the species evolved forward-facing eyes to give them the benefits of stereoscopic vision for more precisely judging distances between thin branches. Their back legs became stronger as they leaped from bough to bough and nails grew on top of their digits exposing sensitive pads at the end of each of them. Their thumbs and big toes began to operate independently of their other digits, giving them the ability to grasp branches and foodstuffs at precarious heights above the ground.

As they evolved such specialties for tree life, so these 'prosimians' (early monkeys) travelled further afield from their place of origin up until some 40 million years ago, at which time they had spread successfully throughout the warm forests of Asia, Europe and North America (only recently severed from Europe by the forces of continental drift).

But in some ways their success was their undoing. One of the species – a tarsier-like prosimian – gave rise to the more powerful monkeys with whom they could not compete. To avoid competition, animals must alter either their life style or their place of living. The prosimians became nocturnal to avoid clashing with the diurnal monkeys. The tarsiers, bushbabies and lorises of today remain nocturnal, but it is interesting to note that some of the lemurs of Madagascar (now the Malagasy Republic) are still diurnal. Significantly there are no monkeys or apes on the island, which was cut off from mainland Africa some 35 million years ago.

The monkeys are not an easy group to understand historically. They are found in Central and South America (part of the New World) and in Africa and Asia (part of the Old World). The two geographical groups look quite similar and as a result they are all lumped together under the popular 'monkey' label. But almost certainly they are only very distantly related and the likenesses of today have been brought about by a natural process of parallel evolution in response to the demands of similar environments.

The New World monkeys have remained almost exclusively in the treetops and some of them have evolved a prehensile tail, which by its ability to grasp branches and even to support the weight of its owner acts as an indispensable fifth limb. So different are these creatures from Old World monkeys that some authorities suggest it is misleading to think of them as monkeys at all.

In the Old World the true monkeys have spread out over a wide area and live in a great variety of habitats. There must have been a time when all of

The large tree shrew (*Tapaia tana*) and its many related species in Southeast Asia are widely regarded as similar to the diminutive tree-dwelling ancestor of monkeys and apes.

these primates were forest dwellers of Africa, Asia and Europe. The turbulent climate changes of about 35 million years ago caused a reduction in these great forests and the creation of large expanses of woodland and grassland. Many monkeys, unable to adapt to the loss of their primary habitat, were forced into extinction, but others – notably the baboons of today – were either not so specialized or were perhaps already living on the edges of the forests. They made the successful move out on to the savannahs where they relinquished their hold on the trees and lived mostly on the ground. In the face of pressure from ground-dwelling predators, they became bigger and developed powerful means of defense. Today the baboons of Africa occupy a great deal of the continent and they have been viewed as presenting useful insights into the behavior of the earliest apes, which much later moved into the woodlands themselves and eventually gave rise to man.

But before this happened, the common ancestor of man and apes was roaming the forests of tropical Africa. Much of the evolutionary story of apes and man is speculative simply because there is such a lack of fossils indicating the steps that the present-day creatures have taken during the last few million years. But at present the evidence indicates the following sequence.

The fortunate discovery in 1960-61 of *Aegyptopithecus zeuxis* in the Fayum Depression of Egypt gives a valuable clue to the evolutionary development of apes and man (see Chapter 5 for the subsequent stages to man). Living

The bushbabies, lemur-like tree-dwelling prosimians from Africa, are regarded as representatives of a stage in primate evolution between the tree shrews and the monkeys and apes.

A spider monkey of South America shows off its prehensile tail, a fifth limb that separates the New World from the Old World monkeys, a group from which the ape line separated some 30 million years ago.

The Evolution of the Apes

When compared to the millions of species that have evolved on Earth (right) since its creation some 4500 million years ago, the ape family (white line) is minute. Their evolutionary significance, however, is out of all proportion to their numbers.

The major living forms – the invertebrates, fishes and plants – were well established some 300 million years ago. Thereafter, the land-based vertebrates diversified steadily, culminating in the reign of the dinosaurs. Only with their extinction at the end of the Cretaceous did the diminutive, nocturnal descendants of the ancient mammal-like reptiles get their chance. In a world empty of large herbivores and carnivores, these early mammals, among them the tree shrew-like ancestor of apes and monkeys, spread rapidly (as did the birds).

The diagram below shows the assumed evolution of the apes and monkeys from the family's common ancestor. The apes are distinguished from monkeys by the flexibility of their arms and their ability to walk upright. Even the gibbon can walk bipedally, though it has to keep its spidery arms in the air to do so. From *Proconsul* (*Dryopithecus*) the lines leading to modern apes diverged. One species, *Ramapithecus*, eventually gave rise to a number of man-like creatures and to *Homo sapiens* himself (see also Chapter 5).

The diagram is by no means definitive, for the apes' fossil record is extremely scanty. This is hardly surprising; the ape family past and present consists in all of some two dozen species spread over a mere 30 million years. By comparison, scientists recognize some 600 species of dinosaurs, which dominated the Earth for 140 million years.

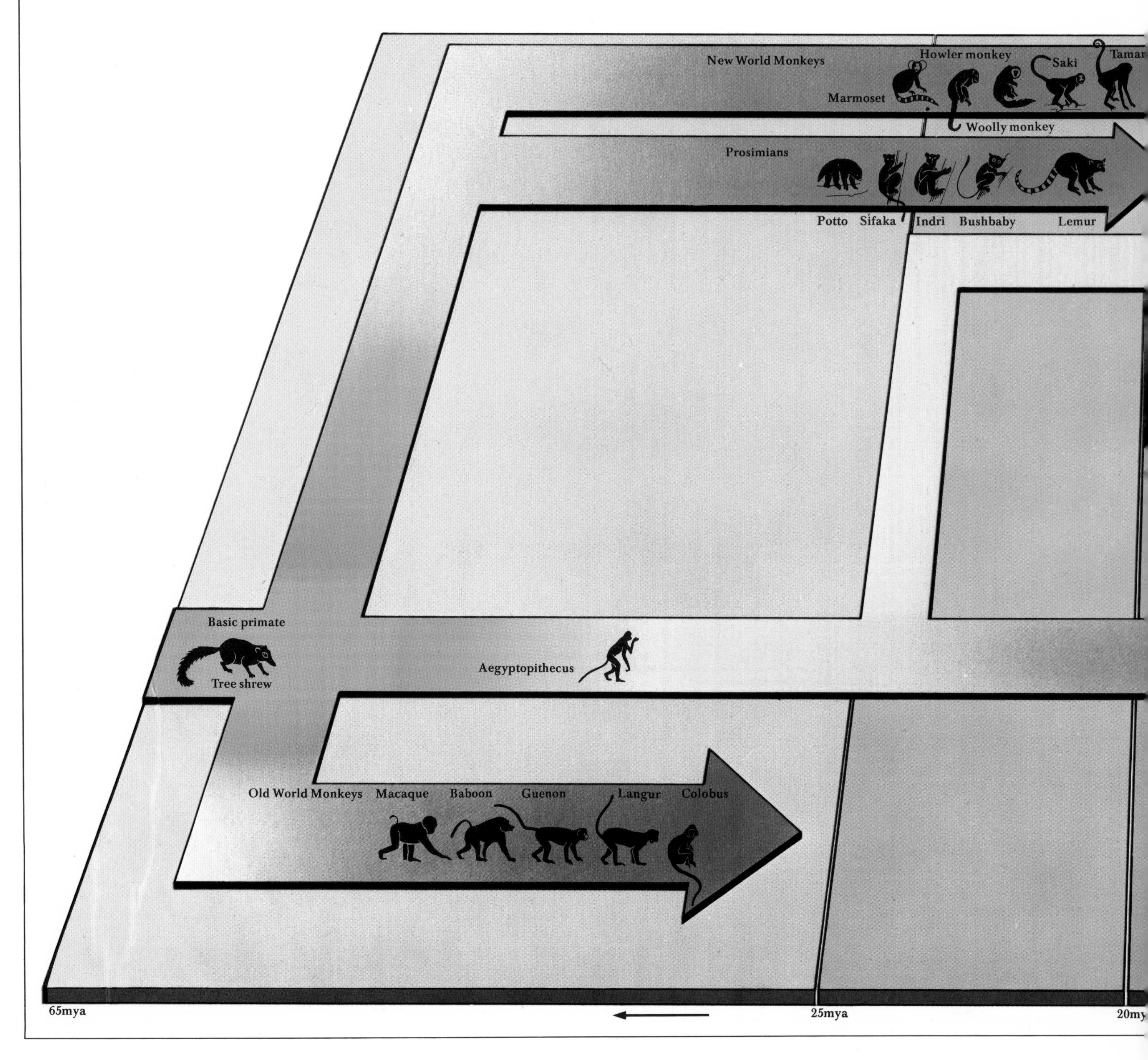

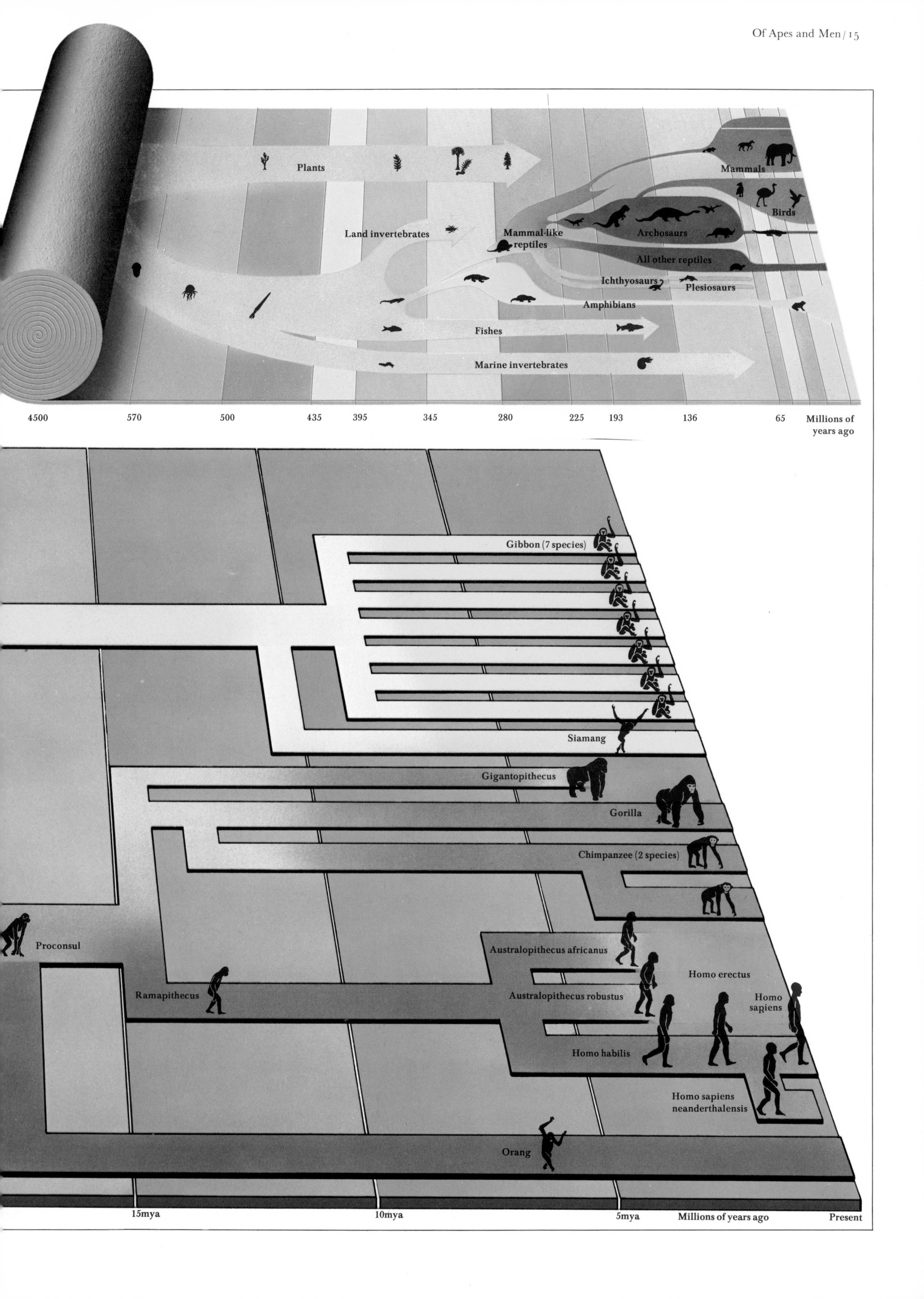
Plants
Land invertebrates
Mammal-like reptiles
Mammals
Birds
Archosaurs
All other reptiles
Ichthyosaurs
Plesiosaurs
Amphibians
Fishes
Marine invertebrates
4500
570
500
435
395
345
280
225
193
136
65
Millions of years ago
Gibbon (7 species)
Siamang
Gigantopithecus
Gorilla
Chimpanzee (2 species)
Proconsul
Australopithecus africanus
Homo erectus
Ramapithecus
Australopithecus robustus
Homo sapiens
Homo habilis
Homo sapiens neanderthalensis
Orang
15mya
10mya
5mya
Millions of years ago
Present

Even the gibbon, seen in this 17th-century German drawing in two versions, was known as a *Waldmensch*, or 'man of the woods' and was accordingly given a human face.

about 28 million years ago, *Aegyptopithecus* showed large canine teeth housed in what looks like an essentially prosimian muzzle, while the remainder of the skull (though the jaw is still missing) followed a more monkey-like pattern. Its teeth followed the ape pattern and out of this mixed bag of characteristics emerged the probability that *Aegyptopithecus* was directly on target for the evolution of the great apes and man. It leads more precisely to *Dryopithecus* (see also Chapter 5), one branch of which gave rise to the more advanced apes. *Dryopithecus major* from Africa is thought to have given rise to the modern gorilla, while the closely related *Sivapithecus indicus* from India may have been ancestral to the orang-utan.

These are very general assumptions because the Dryopithecines were spread across the world and were evolving many different forms over a long period of time. Possibly the relentless pressures of Continental Drift cut off several populations in Africa, Europe and Asia with the result that some went extinct and some became more specialized in restricted areas. This would have forced them to follow different evolutionary lines that could finally have brought them to the modern world as gorillas, chimpanzees and orang-utans.

The living apes so far mentioned are collectively termed the great apes because of their undoubted relevance to the evolution of man. But there is another group – the so-called lesser apes – which, while further removed from the line of our ascendancy, are worthy of consideration because they are actually apes and nothing else. They are the gibbons and the siamangs

SNOWMEN AND BIG-FEET: A LONG-LOST APE?

Since the late 19th century, travellers in the Himalayas have been reporting tales of a huge bear-like or ape-like creature – the Yeti, or 'abominable snowman' – that haunts the higher forests and snowy slopes. There have also been reports of a surprisingly similar creature – the Sasquatch, or 'Big-foot' – which is supposedly a resident of the coastal forests of northwest United States and southwest Canada.

Though the evidence is largely anecdotal, and frequently dismissed by scientists, the legends are extraordinarily persistent, and a number of zoologists remain impressed by them. It is, they point out, theoretically possible for a large creature to remain hidden in remote areas – both the panda and gorilla were mere 'legends' for decades.

Perhaps the strongest evidence of all is that of footprints. In 1951 Eric Shipton, leader of Britain's Everest Reconnaissance Expedition, took photographs of tracks (right and overleaf) that startled the scientific world. Many thousands of similar (but not identical) tracks have been seen across the American Northwest.

The 'Yeti' prints are enigmatic. In Shipton's 'creature,' the toe is not fully opposed as in all living apes, nor is it flat like that of the truly bipedal *Homo sapiens*. Other footprints found in the Pamirs – the northern extension of the Himalayan system in Russia – look very similar to Neanderthal tracks, as do many of the Sasquatch prints.

Other than the footprints, the most dramatic evidence of the existence of one of the creatures, the Sasquatch, is provided by a movie film, of which the picture at left forms a frame. Taken by Roger Patterson near Bluff Creek, California, in 1967, it apparently shows a bear-like or ape-like humanoid, but the sequence is short and blurry and cannot be considered conclusive.

If the creatures exist, what could they be? None of the prints resembles those of bears. There are three remote possibilities within the context of the known evolutionary development of apes. They may constitute relict communities of *Gigantopithecus*, an extinct anthropoid known only from a few fossilized bones and teeth; they may be surviving Neanderthalers (as some Russian scientists believe) driven into remote regions by pressure from *Homo sapiens;* or they may be a giant form of orang-utan, once common on the Asiatic mainland. In any event the possible existence of a species – or perhaps several species – of unidentified biped remains a challenge to science.

from Southeast Asia and they form a family – the *Hylobatidae* – distinct from the *Pongidae* (great apes) and the *Hominidae* (man). All, however, are grouped under the super-family *Hominoidea*.

The gibbons are the best acrobats of all the apes. They swing their bodies through the trees by their arms alone (brachiation) and only infrequently do they descend to the ground. Indeed their arms have become so specialized for this form of locomotion that a gibbon on the forest floor must hold its arms aloft not only to maintain its balance but also to prevent its hands from dragging on the ground. But these same hands which have become so well adapted for carrying the gibbon safely through the treetops at breakneck speeds have inevitably paid the price for such specialization. The extended fingers and the comparatively reduced thumb leave the animal with little ability to grasp objects in its hand.

The earliest gibbon fossils date back some 22 million years and have been found in Kenya and Uganda, indicating their involvement in the general scheme of ape evolution that was taking place at that time and in that region of the world. Like the ancestor of the orang, they spread eastward when the arrangement of the earth's land masses with their continuous tropical forests made it possible for them to do so. Today there are more species of gibbon – six or seven, depending on the classifier – than there are of any other ape. This indicates that they have been adapted to the trees, and have diversified within them, for a very long time.

Because of this life style, they are small in size (never exceeding 30 pounds) and have not been through the processes – especially those exerted by ground-dwelling predators – which demanded an increase in competitive

The Lesser Apes, of which the white-handed gibbon below is a representative, originated in Africa, but are now found only in the forests of Southeast Asia.

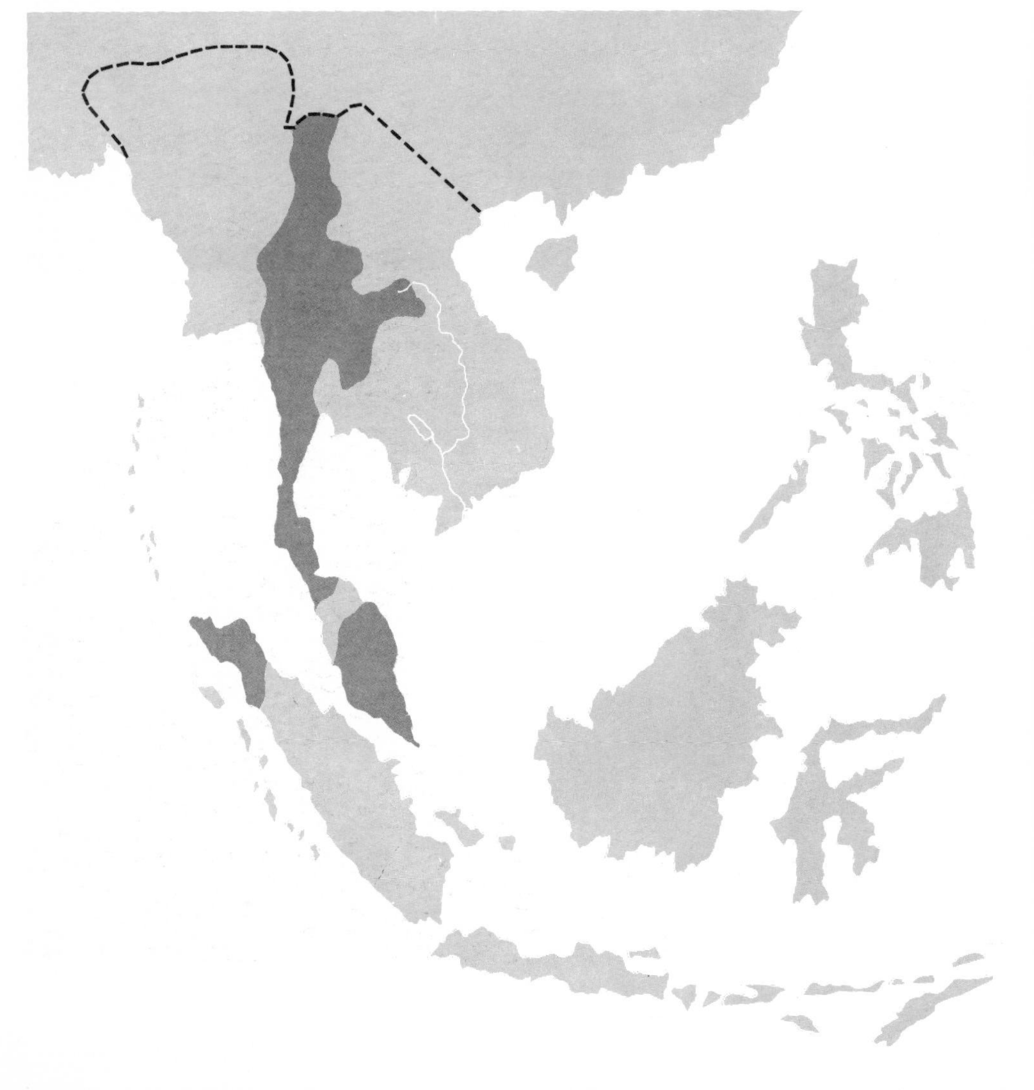

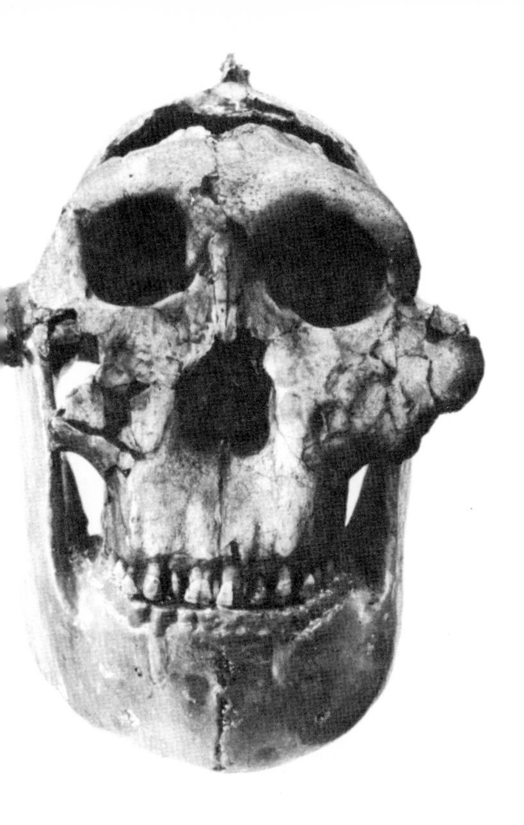

Australopithecus robustus (or *boisei*), a relative of early man, shows the gorilla-like sagittal crest, which anchored the jaw muscles, and the low brain case that probably set the creature at a disadvantage in competition with the brainier *Homo sapiens*.

complexity. They show some great ape trends in intelligence and behavior and yet they retain rather primitive characteristics overall. The siamang probably approaches the great ape grade of organization more closely than does any of the true gibbon species.

The outstanding feature of human evolution has been the development of the brain. By comparison with that evolutionary leap, it is of little human significance that the great apes can be closely allied to us both anatomically and behaviorally, that chimpanzees, for instance, share something like 99 percent of our genetic history. But it is of significance that we talk, dress, write, create and take an active interest in the state of the world around us. No other animal does this. Where did that ability come from? In what ways did a more complex brain aid our survival? The answers to these questions demand an assessment both of our own intelligence and that of apes, and they lead to a surprising conclusion – that we are, almost certainly, not as special in our intelligence as we like to think.

The brain is an expression of genes which gradually spread among the early subhumans. It took time, of course. An organ as sophisticated as the human brain must have been evolving over millions of years. Latex molds from the skulls of early humans show that our brain was more human-like than ape-like at least three million years ago. Allowing 20 years for a generation, that is something like 150,000 generations ago. And at the beginning of that period, the brain was already well advanced. How did it get that way?

This probably is a result of interactions with other parts, e.g. the hand.

A chimpanzee, Bruno, makes the sign for 'listen' in a gestured exchange with researcher Roger Fouts.

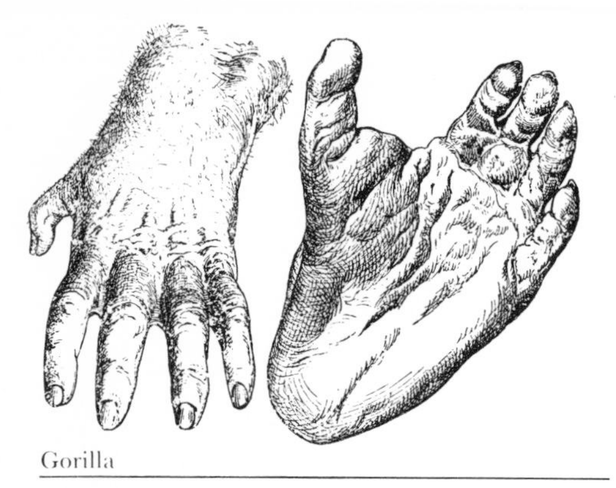

Gorilla

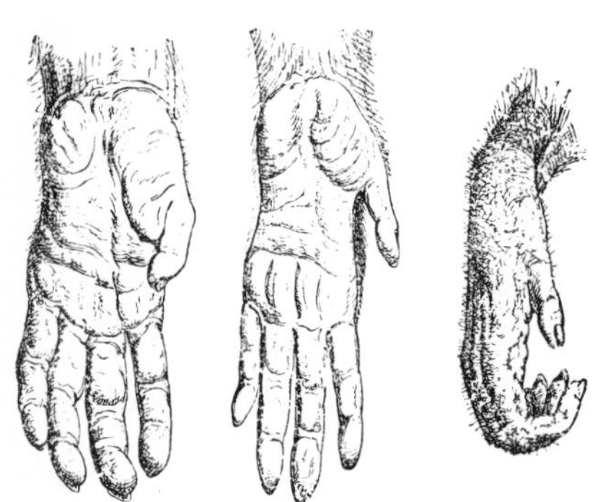

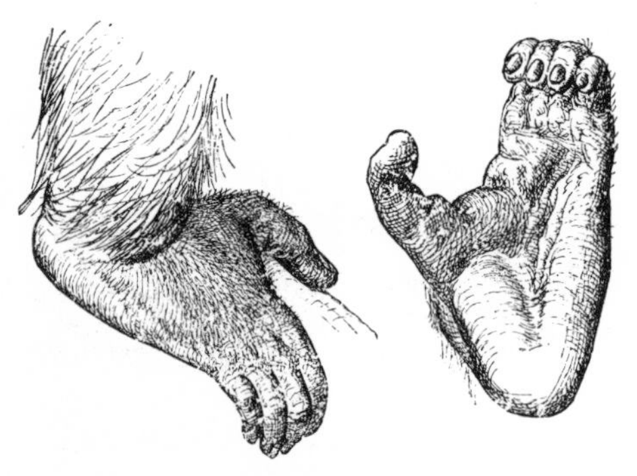

Chimpanzees

Orang-utan

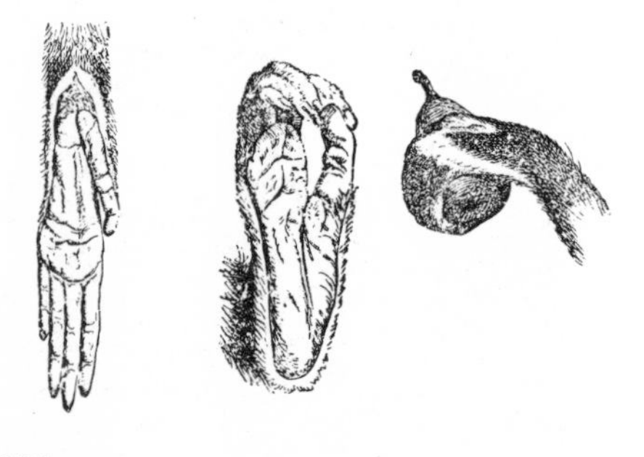

Gibbon

The feet and hands of the three Great Apes, when compared with those of the gibbon, all show opposable thumbs and a flattened structure which allow the animals to walk, climb and manipulate objects with some delicacy.

An analysis of the relationship shows how the brain can become more complex in response to the requirements of a particular organ. There was a time when the ancestral human used its hands more for locomotion – by going around on all fours – than for anything else. But when it sat down to feed, its brain was sufficiently advanced to permit the hand to pick up objects and to place them in its mouth. Gradually bipedalism evolved and the hand became increasingly redundant as an organ of locomotion. When the new gait was fully established, the hands began to develop their own long-term faculties. The fingers evolved to manipulate objects more dexterously. The eyes had to perceive this and send suitable images to the brain. These images had to be understood and if they could be stored for future use, then so much the better. So an enlarged part of the brain dealing with memory would bestow enormous survival benefits upon the individuals so endowed. The facility would soon spread through the population at the expense of those individuals less well endowed and the humans of subsequent generations would find themselves manipulating, perceiving and recalling objects in a way that had not been possible before then.

An ability to reproduce that memory would also have conferred survival, for in this way experience can be passed on second-hand. There is an area of the brain which is called Broca's area. In humans it is larger than in apes and it controls the muscles associated with those parts of the body – principally tongue, lips and larynx – directly concerned with speech. Broca's area is connected by an elaborate system of nerves to the memory-store – Wernicke's area – further to the back of the brain. When these two areas operate together, words can be applied instantly to everyday sensory perceptions. Such an ability would be very useful to a community of ape-like creatures in a hostile world. Survival could result from remembering a dangerous situation which could be recalled and avoided in the future. Benefits would flow to individuals who could utter intelligible sounds which gave added meaning to everyday behavior – and, of course, to those who could understand them. Males must have found the first rudiments of speech useful to communicate more freely when they were out hunting with armfuls of weapons, females as a means of communicating with their children, and children as an aid to playing.

Intelligence, speech, community life, civilization – all are interwoven. It is possible to see how painting – and ultimately writing – could have evolved from the evolutionary foundation. The developing brain, verbal communication, a dexterous hand and the need to relate the vital aspects of the environment combine to suggest the idea of making permanent records. By the time that the first humans were showing even the rudiments of speech, we had left our ape ancestry a long way behind us. Our intelligence and culture – the use of weapons, fire, clothes, storage – allowed us to survive under different conditions. We spread out and away from the forests and moved into regions of the world where no ape would survive for more than a few days, let alone for long enough to breed. We would survive anywhere (even space, as it turns out, given the right artificial environment). We were the world's first generalized species.

While man has spent the last million years or so exploring every aspect of the globe, the apes have been locked away in forests with no genetic option open to them other than specialization. And that is how we find them today – totally dependent upon habitats which we are ourselves reducing daily by frightening amounts. They cannot survive for much longer. [Text continues on page 26]

THE TALKING CHIMPS

Eugene Linden, in 'Apes, Men and Language,' describes the development of the use of sign-language by chimpanzees under Beatrice and R Allen-Gardner, Roger Fouts and others at the University of Oklahoma's Institute of Primate Studies.

'High in the branches of a cottonwood tree on a lushly forested island sit three gibbons. These graceful blond acrobats daily come whistling through the leaves to this spot, assembling like a tribunal of elders to witness and judge the arguments that erupt in a colony of raucous, young chimpanzees on a neighboring island far below. While the gibbon island is verdant with cottonwoods and willows, the neighboring island has only sparse groundcover. A brown African hut used as a communal lodging dominates the center of the chimp island. A fence gives it the air of an abandoned compound. Instead of cottonwoods, there are tall poles where the chimps occasionally perch to spend long periods sitting motionless like figurines. Their gibbon jury, though practiced in observing the unfolding life of the forest, must be confused to see the strange behavior that occurs intermittently between two chimps. One chimpanzee will gesture intricately toward another, perhaps touching his chest and then drawing a finger across the top of his hand. This act would then cause the chimp to which he was gesturing to come forward to tussle and tickle. A keenly observant gibbon might notice that the largest of the young chimpanzees most often resorts to this strange method of communication and that this chimp seems to be a protector of the younger and smaller chimps on the island. This chimp is Washoe. The gestures, of course, are in Ameslan.

Washoe uttered her first combination of words in April 1967, ten months after the start of her language training. She said, "gimme sweet" and later, "come open." She was then between eighteen and twenty-four months old, which is about the age that human infants begin to form two-word combinations.

As Washoe began to combine signs in series such as "you tickle me," the Gardners set about to explore how these two-, three-, four-, and five-word phrases compared with the early utterances of children.

One question was whether or not these combinations were words randomly strung together, or whether they reflected some sense of grammatical order. Most of the doors, closets, and cupboards in the trailer were kept padlocked, which meant that if Washoe wanted to eat or inspect the contents of any of these areas she had to request that they be opened. Fouts and the Gardners noted that she used consistent word order in requesting access to these places. She would say, "open key food" to get in the refrigerator, or "open key clean"

Washoe signs 'bird' to Roger Fouts.

to get at the soap or "open key blanket" in asking to be given a blanket. In requesting people to do things like let her out or hug her, Washoe placed the pronoun "you" before "me" 90 per cent of the time. During this test period, however, 60 per cent of the time she would also place both "you" and "me" before the action verb in phrases such as "you me out"; while 40 per cent of the time "me" would follow the verb, as in, for instance, "You tickle me." What this division represented, says Fouts, was a shift in Washoe's word order that occurred during the testing period, because after this testing period she consistently separated the "you" and the "me" with the action verb. She was shifting toward English grammar in her multiple-sign constructions, a word order preference shared by other chimps presently at the institute in Oklahoma. . . .

Lucy is the oldest of the institute's chimps currently being raised in species isolation. Her foster parents are Maury and Jane Temerlin. Maury is a psychologist who teaches at the University of Oklahoma; Jane is Dr Lemmon's assistant. Lucy was born on January 18, 1966, and removed from her mother four days later, she has been with the Temerlins ever since. Like an institute in miniature, their house is a modern rambling structure of glass and plaster. The east window of the living room looks out on several large wire mesh cages that house a garrulous and excitable crew of blue and white macaws; the south window looks out on a patio and two ponds. When the Temerlins are out, Lucy lives in a spacious indoor-outdoor, wire-enclosed duplex. When they are at home, she lives and sleeps with them.

During my two trips to the institute in the summers of 1972 and 1973, I met Lucy on several occasions and attended a few of Roger Fouts's sessions with her. Fouts and a number of his assistants visit with Lucy one after another for an hour or two each, five days a week. During some sessions, she would be taught new words, during others tested on vocabulary or some aspect of her word usage, while during still others, Lucy and her companion would just talk and review the signs that she knows. Each assistant kept a record of her utterances on a work sheet . . . and described any novel circumstances, difficulties, or errors that occurred during her signing. Imagine how Lucy must feel. There she is, happy just to have a visitor, and the visitor insists on asking her the names of objects that both of them already know. Then when she tries to start a conversation, all the visitor does is sit there and write.

The aim of these worksheets was to build a statistical profile of selected aspects of Lucy's word use, and in this way Fouts has elicited and documented some extraordinary behavior. Lucy's vocabulary is about eighty words. It could be much larger, but the investigators are more interested in the way she uses her vocabulary than in its size; they cannot examine all aspects of language at once.

In focusing on one aspect of language, the investigator must be inattentive to other uses of language that lie outside the thrust of a particular experiment. While the investigator is looking for one aspect of language, Lucy may be gaily demonstrating another. Unlike the investigator, the outsider naturally attends to evidence that the chimp is exploiting the *communicative* aspects of language and is not merely solving problems. And so, as I was introduced to Lucy and the chimps at the institute, I paid particular attention to the flavor of their use of Ameslan, and to those aspects of "speech" that might be obscured by a data table. I found that I reacted to different things than Fouts did. Indeed, the contrasts between Fouts's and my perspectives on Lucy brought out this important characteristic of the investigation of the chimps' use of Ameslan – that the examination of behavior can at times obscure the behavior itself.

This became clear when I met Lucy the next day. I arranged to show up at the

Washoe signing 'time.'

Washoe indicates where she wants to go.

Temerlins a little while after Roger began his morning session. Accordingly, at about 9.30 a.m. I walked around to the patio and peered through the living room window. I could see Roger and Lucy gesturing and cavorting on the couch. Fouts disengaged himself and got up to let me in.

I settled unobtrusively onto a couch to observe and to take notes. Lucy promptly abandoned Roger, hopped onto my lap, and, after some unselfconscious staring, commenced a minute inspection of my face and clothing. She looked at my eyes, peered up my nostrils, and then briefly groomed my hair, presumably looking for lice. I had a scab on my knee, visible because I was wearing tennis shorts, and when Lucy had finally worked down to it, she looked over to Roger and touched the tips of two index fingers together. "She's saying you're hurt," said Roger.

I thanked Lucy, who, chuckling and grimacing, ran back over to Fouts. He showed her a picture of a cat and asked her what it was. "Cat," Lucy replied. For a moment Lucy continued to identify the pictures Roger showed her, but as soon as I picked up my pen to take notes she was dying to see what I was up to, and she raced back over and again hopped onto my lap. As I tried to note this, she grabbed the pen and began to scrawl furiously. Roger pointed out that she was using her right hand (I am left-handed so she was not imitating me) and that she was also holding the pen in a manner similar to a precision grip. Roger also said that Lucy seems to consistently hold objects in her left hand in the power grip. Psychologist Jerome Bruner has observed that in children the dominant hand develops a variety of precision grips while the other plays the role of the steadier, and, to stress the parallel between tool manipulation and language, that the dominant hand plays the role of the predicate to the subordinate hand's subject.

What is significant about Lucy's proclivities in drawing is that right-handedness and left-handedness in humans are related to what is called lateral dominance – the organization and division of labors between the two different hemispheres of the brain. The extraordinary selective pressures that produced language in man required the rapid development of certain parts of the brain, and as a result, rather than both developing equally, the left hemisphere was pushed out of shape to accommodate the renovations necessary to equip man for language. It is possible that the chimp brain is in the first stages of being lopsided. . . .

After Lucy grew tired of drawing ferocious circles, she looked at me and noticed that the white shirt I was wearing had an alligator insignia on it. Lucy pointed to it several times and tracing a question mark in the air asked me what it was. I looked plaintively over to Roger, who suggested that I put my palms together as in prayer and then make the snapping motion of an alligator's jaws. With this advice I laboriously told Lucy that the insignia was an alligator. The chimpanzee cannot flex its hands backwards from the wrist as easily as a person can, and so, when we asked Lucy to identify the insignia, she, after some fumbling, made the sign with the snapping motion originating from the tips of her fingers. It is a testimony to the accuracy that Roger demands of his chimps that he thought Lucy was just babbling and making a confused version of "book," a sign made by unfolding the closed palms in imitation of an opening book. Only after she persisted in her variation did Roger accept that she was attempting to make the snapping sign. This incident made it clear that Fouts was not reading anything into Lucy's signing. It also gave me the feeling that published accounts gave a very conservative and formalistic picture of what the chimps were doing. While humans were peering at the chimps through the lens of experimental design, the chimps themselves were indeed exploiting Ameslan as a means of communication.'

Washoe asks for 'fruit.'

Washoe begs a piggy-back ride.

Before they go, however, scientists are racing against time, particularly in America, to find out how intelligent these relatives of ours really are – research that should cast light not only on the apes but on the evolution of our own abilities.

It is a typically human exercise and it began with the earliest awareness of the similarities between apes and man. The natives of Africa and South-east Asia believed that the apes around them were humans who had blasphemed in some fearful way and had been cast out to live in shame in the forests. The most horrific stories of violence were pinned upon their names and it is not surprising that the early European explorers, often overready to emphasize their own daring, were only too eager in their ignorance of the ways of evolution to perpetuate such myths.

The story of how these myths have been largely dispelled by the actions of just a handful of dedicated scientists is a fascinating one and it is traced in later chapters of this book. The more recent progress which has been made in understanding how closely these apes approach the human level of intelliegence is another matter.

It has long been accepted that speech is an unbridgeable gulf separating humans from all other animals. We have come to look upon it as representing some divine gift. Those of us who are alive today can have no idea of the way in which language was acquired. How significant is the gulf of language? And how does it relate to intelligence generally? Perhaps the apes can tell us.

The fossil records give no real clues as to where our intelligence came from and in attempts to understand this better many experiments have been carried out among different kinds of animals. Chimpanzees do very well in these tests, as befits an animal resembling us in so many ways. But still they do not talk and their ability to copy almost everything else that we do endorses further the idea of speech being a special prerogative of mankind.

Still is it fair to conclude that chimpanzees are stupid creatures because they do not talk? Obviously not. They do not talk because they were never in a situation in which it was necessary to evolve the mechanisms for so

Yula, a young chimp in London Zoo, displays the facial mobility that, when suitably modified gives chimps a range of expression to rival our own.

THE APE AS ASTRONAUT

Because of their similarities to man, apes are invaluable substitutes in scientific research. Without them the first astronauts would have faced far greater risks. In 1949 a rhesus monkey was rocketed 83 miles high. Subsequent tests included the 1961 suborbital flight of the chimpanzee Ham, and in all some 75 chimps were accustomed to space equipment.

One of the questions that the test chimps helped research was what man should breathe in space. In the early 1960's, after testing the effects of various gas mixtures on monkeys, scientists settled on about one-third normal atmospheric pressure. But pure oxygen is highly inflammable, as was tragically shown on February 21, 1967, when three astronauts died in a fire in their Apollo training capsule. Later a combination of oxygen and nitrogen was specified for extended missions.

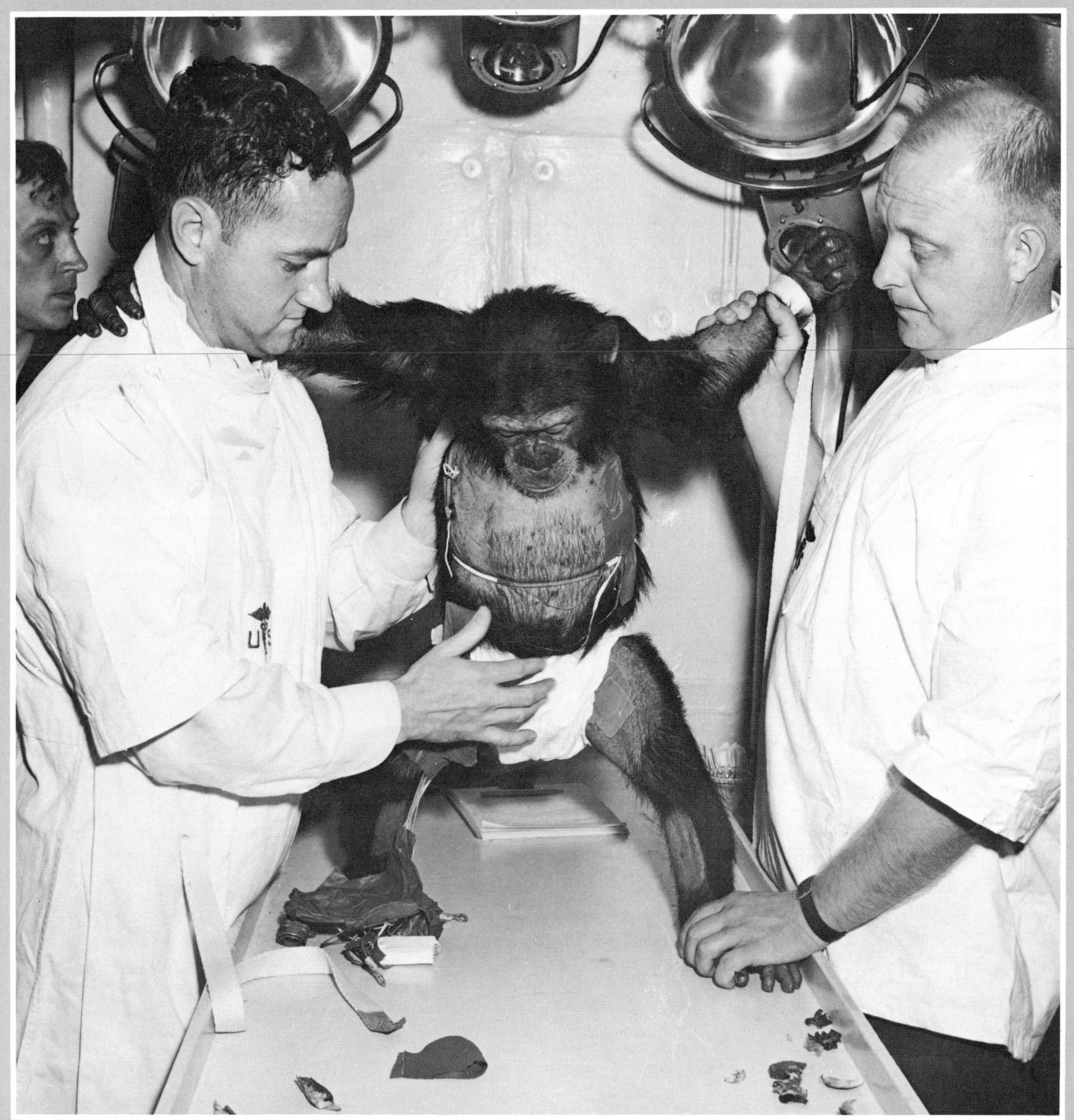

Ham receives a physical after a 5000-mph Mercury-Redstone rocket flight.

Enos, a 5-year-old chimp, gives a rebellious look at the camera from a 'conditioning couch.'

Ham, the first 'chimponaut,' continues his training.

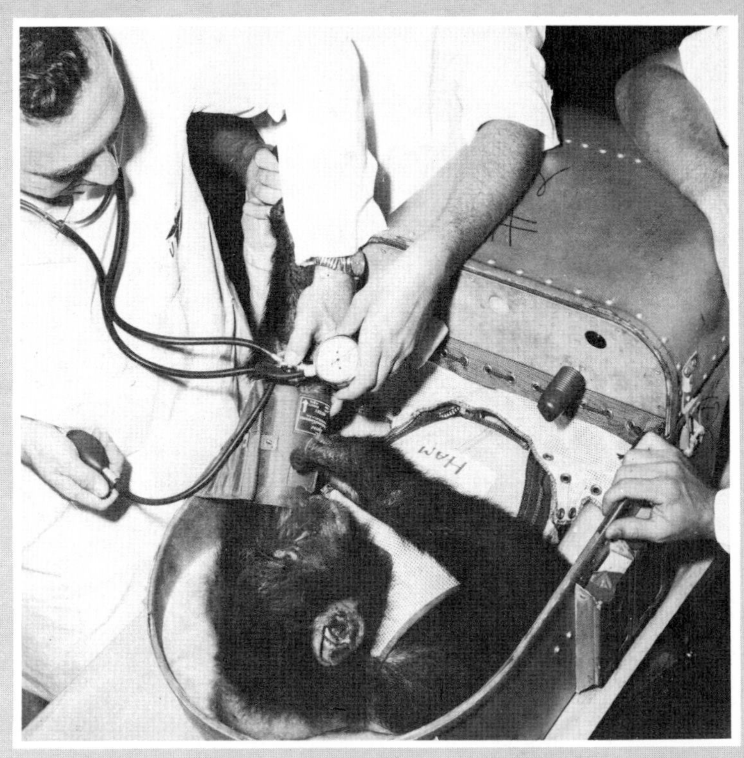

Ham in space suit and simulated capsule.

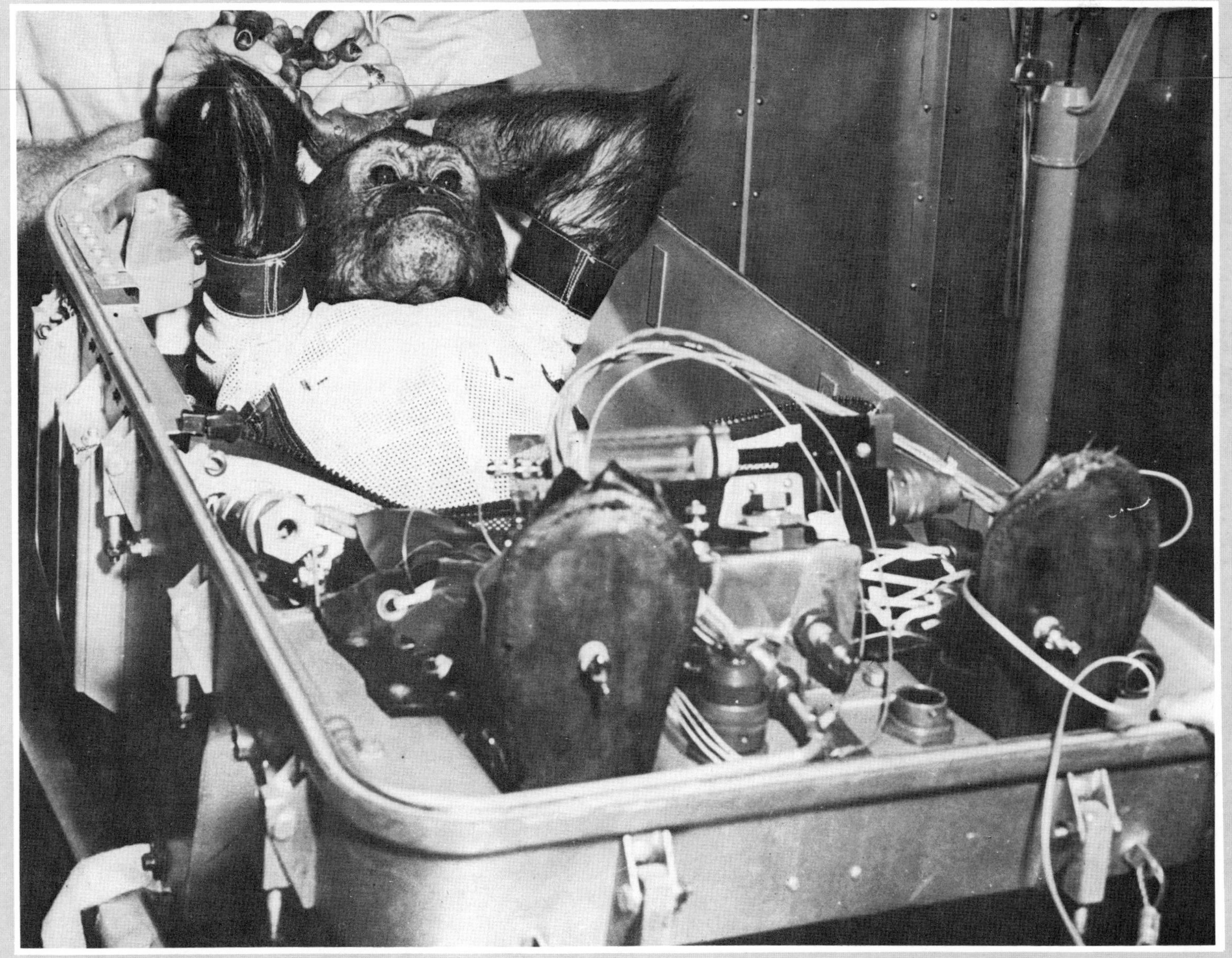
Enos patiently awaits an orbital launch from Cape Canaveral atop a Mercury-Atlas rocket.

doing. They may well have been prevented from ever reaching that situation by the competition of early humans. Even though they have come through their own processes of evolution since that time, we might expect them to retain the potential that was undoubtedly possessed by the common ancestor of apes and man. This suggests a rough-and-ready argument: if a chimpanzee is an almost-human animal, then with a bit of honest brow-beating it should be able to learn to talk.

In the 1950s the first serious attempt to teach a chimpanzee to talk was undertaken in America by Cathy and Keith Hayes. After six years Vicki – as she was named – showed a childlike ability to sort colors and retrieve balls by pulling certain strings, but she could utter no more than four human words – 'up,' 'cup,' 'mamma,' and 'papa.' By the early 1960s, it was generally accepted that chimpanzees could not talk because their brain was not sufficiently advanced.

But it is not the nature of scientists to let sleeping dogs lie. In about 1920 Robert Yerkes had suggested that some sort of gestural language might be the most satisfactory way of establishing a communicative relationship with chimpanzees. It made sense to try to talk in the chimps' language rather than force them to talk in ours. The suggestion was taken up by Beatrice and Allen Gardner. They worked on the simple understanding that language is not necessarily restricted to speech. Chimpanzees do not have the vocal apparatus to formulate sounds as we do and this was the reason for the failure of Vicki to augment significantly the grunts and hoots she was born to utter. The Gardners' approach was realistically based on the anatomical peculiarities of their subject, and it presupposed nothing about the quality of the brain. Above all it sought the best medium through which the brain could express itself in ways readily understandable to humans.

In 1966 the Gardners acquired their first chimpanzee from the wilds of Africa – a one-year-old female who had been separated from her mother.

Aping Humans: An Appealing Caricature

In zoos the extraordinary similarities between chimps and humans have for years been exploited in 'chimps' tea-parties' to provide entertaining caricatures of human activities. One English company, Brooke Bond, has taken this idea to its logical conclusion by persuading zoo chimps to act out 'slice-of-life' sequences as advertisements for their PG Tips tea. The series has been phenomenally successful: the first chimp ad appeared the night commercial TV began in 1956 and since then some 60 films have been made. The ads are hard to make – only about one foot of film is shown for every 100 feet shot – but the crew use two cameras to limit time spent on set and after work the 'actors' happily accept their rewards – fruit, soft drinks, and, of course, cups of tea.

She was named Washoe after the county in Nevada where she spent so many of her early years. She was given an environment rich in objects of curiosity, including all manner of games and toys within spacious living and playing areas. She did not see another chimpanzee for at least six years and the humans around her during this time, particularly Roger Fouts who played such an important role in her education, never spoke in her presence. Instead they conversed in Ameslan (American sign language for the deaf). Ameslan, with its basic 55 signal units, was to be Washoe's way of communicating across the species barrier.

First she had to be made aware of the importance of her hands in a way that was completely alien to her nature. Not only were they useful in manipulating objects but they now had to be used also for manipulating words in a meaningful sense. She took to the idea very quickly, soon learning that it was not enough to run across a room and thump the door when she wanted to go outside. She had been taught the proper signs and although she initially made them when she was at the door, the day came when she signalled her requirements beforehand from elsewhere.

Within three years Washoe had learned 85 different signs, and she used them not only individually but also in combinations of two and three. The first combinations materialized before the first year of her training had elapsed and when she was at about the age when young human beings string their first words together. Her repeated use of correct sequences of words demonstrated how well she could sort them out in her mind.

Beyond any shadow of doubt Washoe and those chimpanzees who have followed her to the Ameslan centers have proved that what goes on in a chimpanzee's brain is something more than it can express vocally. The quality of Washoe's brain must rate very highly. Not only could she identify objects and construct simple sentences but she could do so in the 'language' of another species. Even now Ameslan-proficient chimpanzees are teaching

Apes' faces have a character that is as real as that of human faces. Individuals are easily recognized by observers, as are a variety of emotions – curiosity, attention, excitement, fear, affection, anger, joy and sadness all find expression, as this selection of portraits suggest.

Chimpanzees have considerable artistic skills, as Jenny, a circus chimp, shows here in a Munich school of painting. Some chimpanzees display a rudimentary sense of composition and a desire to experiment with new techniques. In 1964 the creations of one chimp, Pierre, were exhibited anonymously in Sweden and several were given serious critical comment and sold before the hoax was discovered.

members of their own species to communicate between themselves in sign language.

The implications of this minor revolution in the animal world have not yet been fully understood, largely because there is no evidence to suggest that chimpanzees communicate on such a sophisticated level in the wilds of Africa. They obviously carry the potential to do so, but it may be that their intellect is channelled off into maintaining social and sexual stability within their natural groups. The almost inevitable conclusion is that they think about things far more than they indicate by their behavior.

One problem with saying that chimps have a language ability is a proper definition of the word 'language' itself. At what stage of life is it acquired? Is it defined by the understanding of a single word or a single sentence? Is it more than that? Or less – no more than a generalized talent? This is difficult enough when talking about humans, but a chimpanzee that can name objects, signal its intentions and even swear a little, all without 'talking,' demands a rethink of what we mean by 'language.'

Pursuing completely different evolutionary pathways for this great length of time, the great apes and man arrive together on the threshold of the 21st century. The extrovert qualities of the chimpanzee have helped it to participate in experiments which are beginning to reveal how closely related we all are. If only we could devise similar tests for the gorilla and the orang – so solitary, introverted and unco-operative by nature – we might one day bring them out of their thinking world to perform feats that would leave few doubts as to the evolution of our own intellect.

This laboratory work may seem a far cry from the plight of the individuals still living in the vanishing wilds of Africa and Southeast Asia. But it is not. Those individual apes are our close evolutionary relatives. They have an intelligence we *ought* – on moral grounds alone – to preserve and learn from. Is it right that they should be made extinct as a consequence of our own success? We have guaranteed our own survival through cultural development. Surely we can do no less for our ape relatives?

GIBBONS: THE JUNGLE ACROBATS

The ability to brachiate, or swing hand over hand through the trees, is a characteristic of the apes, although not all of them make frequent use of it. The undisputed masters of the art are the gibbons and siamangs whose slight bodies, elongated arms and hooked fingers make them beautifully suited to arboreal locomotion. The wrist, arm and shoulder are especially adapted for mobility as the animal reaches, grasps and changes its holds. When a gibbon is peacefully brachiating along, it usually travels at about the speed of the average human walk. But when it is excited or frightened, it can plunge through the forest canopy at astonishing speeds, sometimes covering 30 feet or more in a single jump.

A white-handed gibbon takes off in a tree-to-tree leap.

A siamang on a jungle path holds its long arms aloft to prevent them dragging.

In its proper domain, a Sumatran siamang (*Symphalangus syndactylus*) brachiates.

Two white-handed gibbons. Members of this species are often dark, but coloration varies from individual to individual and the identification of species is often hard.

White-handed gibbons in both light and dark coloration.

White-handed gibbon brachiating (above) and showing the species' characteristic face fringe (below).

2/ORANG-UTANS: THE RED APES

The orang-utan – the 'wild man of the woods' of Southeast Asia – is now restricted to three pockets of forest in Borneo and Sumatra (below), although once it was distributed on the Asiatic mainland. Living amidst lush vegetation, orangs seldom exert themselves more than is necessary and have always been easy victims for man. They are therefore well known in captivity, where their humanoid positions (left) and their 'jowly' obesity make them instantly appealing. But only recently have researchers revealed the details of their daily life in the wild.

The orang-utan is a great red ape cast away on the islands of Borneo and Sumatra thousands of miles from the African home of its ancestors. Significantly the orang's isolation has been its salvation. Spreading eastward from Africa at a time when luxurious forest growth catered for its needs, the orang must have been common on mainland Asia some 500,000 years ago, during the Pleistocene era. Fossil remains suggest that those orangs were much larger than contemporary ones and were probably ground-dwelling, roaming woodland and forest alike in groups akin to those of the gorilla today.

In Asia they fell victims to the last great Ice Age which began some 100,000 years ago, locking up the water of the world's oceans in polar ice caps and exposing land previously under shallow water. Advancing cold forced them southward from China to find refuge in the tropical forests of an area then connected tenuously to the main bulk of land further north. Here, at the end of the Ice Age, they were cut off as invading seas flooded the land bridge. The islands of Borneo, Sumatra and Java (among others) were formed and although the mainland climate was to improve considerably, the orangs were isolated, unable to retrace their evolutionary footsteps.

They were then forced to change their behavior, for they found luscious fruits and safety from predators high up in the trees that covered much of the islands. With a pressure to shift so completely from a terrestrial life to an arboreal one, natural selection favored the smaller, lighter individuals whose chances of survival to sexual maturity were consequently enhanced. As these individuals reproduced, so the genes controlling their physical stature found expression in the bodies of their offspring. As a result orangs have become much smaller, perhaps only half the size of their mainland relatives of the past.

Tucked safely away from the mainstream of human evolution, orangs flourished at a time when Stone Age man would surely have exterminated them all on mainland Asia. Eventually man spread into Indonesia and began to clear the forests from low-lying grounds, forcing the wildlife high up onto the mountain slopes if it was to survive. On accessible Java even the highlands came under attack and there the orangs died out, easy game for hunters equipped with spear and blowpipe.

But in Borneo and parts of Sumatra the great apes survived. These islands, especially the former, were more difficult to reach and those who made the journey were presented with hostile coastlines that argued against landing. Where early man *was* able to gain a footing, he could do so only tenuously, for inland the green-clothed mountains provided too great a barrier to rapid colonization.

But as the centuries passed by, the human population spread. There is evidence to suggest that the two species lived in close association for thousands of years before the first Europeans arrived on the scene. Perhaps the forest natives were well aware of the dangers of over-exploiting their environment and managed to maintain a balance, vital to their own existence, which involved the continued existence of the orangs. Perhaps also they were suspicious of them, recognizing their undoubted human qualities, but still fearful.

Whatever the truth, orangs were part of their daily lives. The word 'orang' is of Malay origin. It means simply 'man' and was once used to identify natives from the great inland forests of many of the Indonesian islands. More specifically it means 'rational being' and as such was often used as a term of reverence both for tribal chiefs and for their chief animal,

These two drawings from the 18th century (top) and the 19th century (bottom), though separated by over a hundred years, emphasize the orang's legendary 'man of the woods' image. Both show them as sage-like creatures holding staves. Orangs do not carry sticks. The idea probably arose because they do break off branches and use them as missiles, and because they normally find it difficult to stand upright – and use their arms as crutches.

the elephant. Finally the word was also applied to the old ape which lived peacefully in the forests around the villages. The word 'utan' ('utang' – 'in debt' – is a wrong spelling) means 'of the woods' and the joining of the two words to form orang-utan distinguished the non-human ape from any other creature in the area.

But local dialect differences, lack of communication and a host of superstitions must have led to considerable confusion as to who was whom and what was what. The first European visitors were certainly confused. When a Dutchman, Dr Jakob Bontius, arrived in Indonesia during the first half of the 17th century, he related in 1630 how he had seen orangs walking upright on two legs and how the females clutched their private parts with obvious embarrassment in the presence of a stranger. Orangs groaned and cried in the manner of man, he said, but did not actually speak. We can do no more than speculate as to whether he was trying to describe an orang or whether he was simply combining men and apes in his imagination.

In 1641 Europeans first applied the name orang-utan to a great ape – but to the wrong one. The ape in question came from Africa – it was a chimpanzee that had been captured in Angola. The misnomer was perpetrated by an Oxford don, Edward Tyson, a member of the Royal Asiatic Society, whose chimpanzee unfortunately died *en route* from Africa to England. He returned to London empty-handed, swearing that he had attempted to bring a live 'orang-utan' back to Europe with him.

It was not until 1718 that the ape was rightly described under its proper name – something over 200 years after the first European set foot in Borneo. Captain Daniel Beeckman, an English sailor, drew up some interesting accounts of orangs in which he confirmed that the natives really did believe that these were formerly men but were metamorphosed into beasts for blasphemy.

Clearly Beeckman did not believe this himself, for he gave a fair appraisal of the grades of organization within the closely related primates of the area: 'There are many kinds of monkeys, apes and baboons [there are no true baboons in Borneo, though there are many other types of monkey, some species of which were loosely known as baboons] which all diverge greatly from one another in form. The most remarkable among them are the orang-utans. They are up to six feet tall, walk upright on their feet, have longer arms than man, and possess a tolerable face in appearance. They seemed more attractive to me than many a Hottentot that I have seen. Furthermore they have large teeth, no tails, and have hair only in those places where man also has hair. They move easily and have tremendous strength. They throw stones, sticks, and pieces of wood at people when they feel that they are being attacked.' This may be a slightly anthropomorphic and fanciful description of orangs but it remains the earliest reasonably accurate one.

Beeckman, in fact, was so possessed of the charms of the apes that he purchased a youngster to keep on board his ship. An extract from his diary reveals what an engaging time it must have been for them both. This ape proved to be 'a great thief and loved strong liquors; for if our backs were turned, he would be at the punch bowl, and very often would open the brandy-case, take out a bottle, drink plentifully, and put it very carefully into its place again.'

The natural problems presented by Borneo – its lack of suitable harbors and its massive, rugged inland regions – remained to repel European explorers until well into the 19th century. In 1839 a young British globetrotter, James Brooke, founder of Singapore 20 years previously, stopped off in

This hand-colored lithograph from the mid-19th century groups the orang along with monkeys from Africa. Monkeys and apes are only distantly related; the print reflects a pre-Darwinian ignorance of the creatures' true evolutionary status.

Brunei (a province of Borneo) to find himself the recipient of an extraordinary offer. The Sultan, learning that Brooke was well armed, asked for help in suppressing some rebels. In return he would make Brooke a Rajah and give him Sarawak to rule. Brooke's thirst for adventure – and, no doubt, thoughts of power and riches beyond his wildest dreams – led him to accept the offer. Down went the rebels and Brooke was duly installed as the first white Rajah of Sarawak (a dynasty that lasted until 1941).

For all the pomp and ceremony attached to his elevated status in life, Brooke had a thankless task on his hands. When he arrived the local Dyak tribes were being harassed by Malay traders and chiefs. They suffered brutally: their women and children were captured and sold as slaves, their villages were plundered and neighboring tribes took advantage of their misfortunes to attack them. Brooke arrived as their salvation and under his careful policies the Dyaks were able to live in peace. It is a reflection of the man that the Malay people against whom his harshest strategies were directed also came to respect him.

As well as establishing his rule, Brooke found time both to hunt and study the wildlife around him. He was particularly interested in orangs. The first male orang he killed excited him so much that he wrote to the Zoological Society of London, 'Great was our triumph as we gazed on the huge animal dead at our feet, and proud were we of having shot the first orang we had seen, and shot him in his native woods, in a Borneo forest hitherto untrodden by European feet.' But generally the new Rajah considered orangs to be boring and lethargic, for they never bothered to overexert themselves when being pursued and he could always find them again, even if he had to wade through a swamp up to his neck in pursuit. Brooke was convinced they would wait for his straggling party to catch up with them – not much sport for a hunter.

Brooke also had a scientific interest in his prey. He had met and befriended Alfred Russel Wallace, the eminent professional collector who arrived in Sarawak in 1854. Apparently the two men conversed frequently on science and religion and Brooke soon became fascinated by the ideas of the man who was destined to become – along with Charles Darwin – the co-founder of the theory of evolution. Whenever he could, Brooke sent details of orangs to Wallace. The two men were slightly at odds in their sentiments and appreciation of zoological matters, as the following extract from Wallace illustrates:

'In a letter from Sir James Brooke, dated October 1857, in which he acknowledges the receipt of my Papers on the Orang, published in the *Annals and Magazine of Natural History*, he sends me the measurements of a specimen killed by his nephew, which I will give exactly as I received it: "September 3rd, 1857, killed female Orang-utan. Height, from head to heel, 4 feet 6 inches. Stretch from fingers to fingers across body, 6 feet 1 inch. Breadth of face, including callosities, 11 inches." '

'Now, in these dimensions, there is palpably one error,' continued the rigorous Wallace, 'for in every Orang yet measured by any naturalist, an expanse of arms of 6 feet 1 inch corresponds to a height of about 3 feet 6 inches while the largest specimens of 4 feet to 4 feet 2 inches high, always have the extended arms as much as 7 feet 3 inches to 7 feet 8 inches. It is, in fact, one of the characters of the genus to have the arms so long that an animal standing nearly erect can rest its fingers on the ground. A height of 4 feet 6 inches would therefore require a stretch of arms of at least 8 feet! If it were only 6 feet to that height, as given in the directions quoted, the

James Brooke (top), the founder of Singapore and Alfred Russel Wallace (bottom), the co-originator with Charles Darwin of the theory of evolution by natural selection, were among the first zoologists to collect detailed information of orangs in the wild.
The National Gallery

HANDS AND FEET FOR JUNGLE TRAVEL

The most arboreal of the apes – the gibbons and the orangs – have specialized hands with extraordinarily long, strong fingers that can be fixed into hooks for hanging and swinging.

The thumbs of orangs and gibbons however, are different. That of the orang (below) is stumpy and does not get in the way during brachiation (progression by swinging). The gibbon's thumb is longer, and although sometimes used in climbing it can be neatly tucked alongside the palm when its owner is using its other fingers to travel by swinging through the trees from branch to branch.

Although the heavier orangs are fully as arboreal as gibbons, they are as slow and deliberate as gibbons are rapid and daring. In the trees they normally move with great care, testing branches and distributing their weight on as many as possible on the chance that one might break.

When they are frightened or disturbed, however, they forget their caution. At such times, they have been seen to descend a tree by allowing themselves to fall, fleetingly gripping a branch with a hand or foot as they descend. When they are at ease, they hang from a branch by their feet, but with their bodies upright, their legs reaching up on either side above their heads, and their hands resting on their bellies.

The hands are not ideal for plucking fruit, for the thumbs are reduced in response to the need to climb. But they are strong enough to rip off branches to build nests and, when necessary, to throw them at intruders.

On the ground, orangs walk very awkwardly; although they go on all fours, their arms are much longer than their legs, and as a result their bodies are raised up as they move, giving them the look of an old man, bent by age and making his way with the aid of two sticks. Both hands and feet are so curled that they cannot walk with their soles flat on the ground but rather must walk on the outside edges of their feet.

As these two pictures show, the orang is able to move its lower limbs with remarkable flexibility, for it lacks a ligament which connects the femur to the hip socket.

animal would not be an Orang at all, but a new genus of apes, differing materially in habits and mode of progression.'

Wallace goes on to quote accepted examples of error in size estimation, especially as a result of measuring bloated individuals long after they have been killed, and concludes that 'On the whole, therefore, I think it will be allowed that up to this time we have not the least reliable evidence of the existence of Orangs in Borneo more than 4 feet 2 inches high.' (As it happens, this is something of an underestimate. Today the length of head and body is commonly quoted as from 4 feet $1\frac{1}{2}$ inches to 4 feet $9\frac{1}{4}$ inches, with an exceptional male from Borneo reported at 5 feet 11 inches.)

In March 1855 Wallace travelled to the Simunjon coal-works, situated on a small tributary of the Sadong River east of Sarawak. He says that he was happier here than at any other time during his 12 years collecting in the tropics. Here in particular he gained a great deal of first-hand information on orangs. Although in present-day terms he could be accused of shooting far too many of them it was his job as a scientist to secure scientific specimens for Europe and to resolve the varied opinions of the numbers of species. He knew that only a large collection of different ages and sexes from different areas would offer a solution. He was not, after all, confronted with the conservation problems we are faced with today and was collecting at a time when animals seemed to abound almost endlessly in little known regions. Like almost all his colleagues, he saw no conflict between genuine interest and a readiness to kill, as the following incident shows.

These two 19th-century drawings show how the orang can be turned from an easy victim to an aggressor by the activities of man. Aloft, orangs make an easy target (top). If attacked on the ground they can defend themselves with some ferocity – giving rise to the myth that they were a naturally aggressive species, as indicated by the somewhat fanciful drawing at right.

He was out one day observing three small orangs near his camp at Simunjon. He chased them specifically to see how they made their way through the trees at such a tender age and was fascinated by their youthful agility. 'We had a long chase after them, and had a good opportunity of seeing how they make their way from tree to tree, by always choosing those limbs whose branches are intermingled with those of some other tree, and then grasping several of the smaller twigs together before they venture to swing themselves across. Yet they do this so quickly and certainly, that they make way among the trees at the rate of full five or six miles an hour, as we had continually to run to keep up with them.'

In itself this is a worthy piece of observation, with its keen eye for the kinds of branches which dictated the route taken, the way the various twigs were held and, especially the speed at which the youngsters travelled. But note the sequel: 'One of these we shot and killed, but it remained high up in the fork of a tree; and, as young animals are of comparatively little interest, I did not have the tree cut down to get at it.'

Wallace was not a very good shot. By his own admission he shot only one orang with a single bullet and often he had to fire many times before his specimen succumbed. The details of such haphazard collecting make gory reading, as another extract shows:

'Two shots caused this animal to lose his hold but he hung on for a considerable time by one hand, and then fell flat on his face and was half buried in the swamp. For several minutes he lay groaning and panting, while we stood close round, expecting every breath to be his last. Suddenly, however, by a violent effort he raised himself up, causing us all to step back a yard or two, when, standing nearly erect he caught hold of a small tree, and began to ascend it. Another shot through the back caused him to fall down dead....'

He wrote of yet another incident:

'As soon as I had fired, he moved higher up in the tree, and while he was

doing so, I fired again; and we then saw that one arm was broken. He had now reached the very highest part of an immense tree, and immediately began breaking off boughs all around and laying them across and across to make a nest. It was very interesting to see how well he had chosen his place, and how rapidly he stretched out his unwounded arm in every direction, breaking off good-sized boughs with the greatest of ease . . . so that in a matter of moments he had formed a compact mass of foliage which entirely concealed him from our sight . . . I therefore fired again several times, in hopes of making him leave his nest; but, though I felt sure I had hit him, as at each shot he moved a little, he would not go away. At length, he raised himself up so that half his body was visible, his head alone remaining on the edge of the nest. I now felt sure he was dead.'

Wallace provided the first detailed account of orang behavior. Much of it, not surprisingly, was wrong.

He decided that the 'Mias' (as he called the orang, from a local name) descended to the ground only rarely – when pressed to find sufficient food in the trees and during very dry weather when its bodily needs for water, usually satisfied by an intake of succulent fruit, were not catered for.

In fact orangs spend much more time beneath the trees than Wallace ever supposed, for males will come down and forage for hours at a time and will even sleep on the forest floor. It seems logical that as they age these large-sized males are denied access to the treetop routes that supported their weight during preceding years. An inspection of museum specimens soon reveals the number of mended bones that testify to the tumbles they must suffer. As they increase in weight and spend more time on the ground, so they are more likely to come into contact with natives, many of whose stories, applied universally, must concern aged and short-tempered individuals.

When they travel at ground level, orangs do so on all fours. Wallace saw two young ones playing together on the ground at Simunjon and he noticed that they stood erect only when grasping hold of each other. He was convinced that orangs could not walk upright unless they supported themselves by clinging to overhead branches, although when sufficiently angered to attack humans they could sustain a short two-legged charge! Perhaps, once committed to a bipedal defense of its life, the unfortunate animal has to find something to cling on to before it loses its balance and topples over, hopelessly exposed to the hunter's weapons. All the early engravings of orangs attacking natives show them locked in hand-to-hand combat and I wonder if the surprised ape is not really trying to stand up straight and is being murdered more as a result of its anatomical peculiarities than its devilish intent!

Wallace also quoted Dyak sources for the fact that orangs had two natural enemies (besides man): the crocodile and the python. One chief, who swore to have witnessed such an encounter, informed him that 'the only creature he ever fights with is the crocodile. When there is no fruit in the jungle, he goes to seek food on the banks of the river, where there are plenty of young shoots that he likes, and the fruit that grows close to the water. There the crocodile sometimes tries to seize him, but the Mias gets upon him, and beats him with his hands and feet, and tears him and kills him.'

Another chief told him, 'No animal dare attack it but the crocodile and the python. He always kills the crocodile by main strength, standing upon it, pulling open its jaws, and ripping up its throat. If a python attacks a

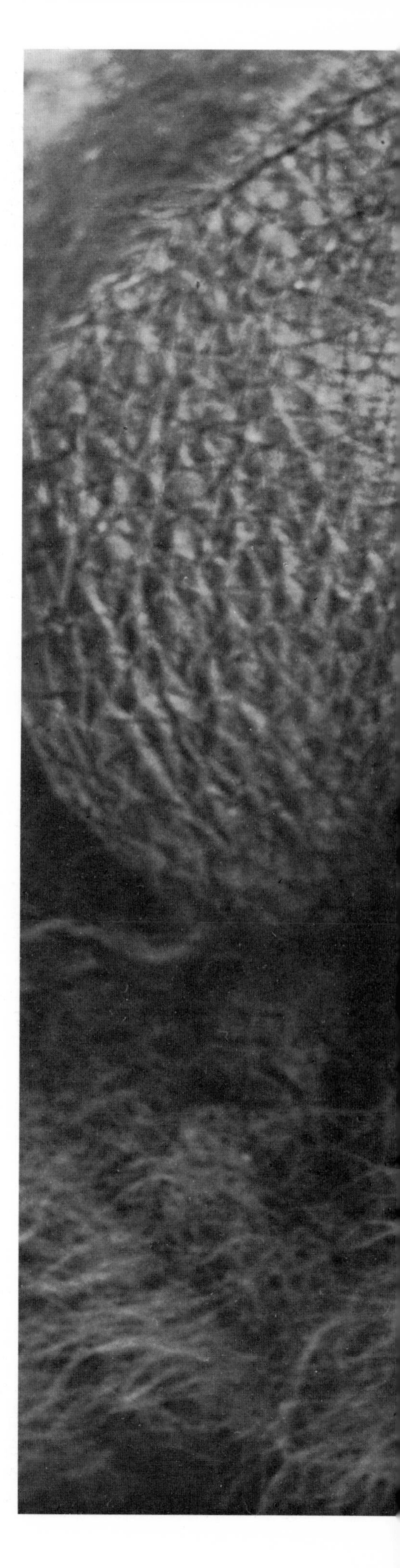

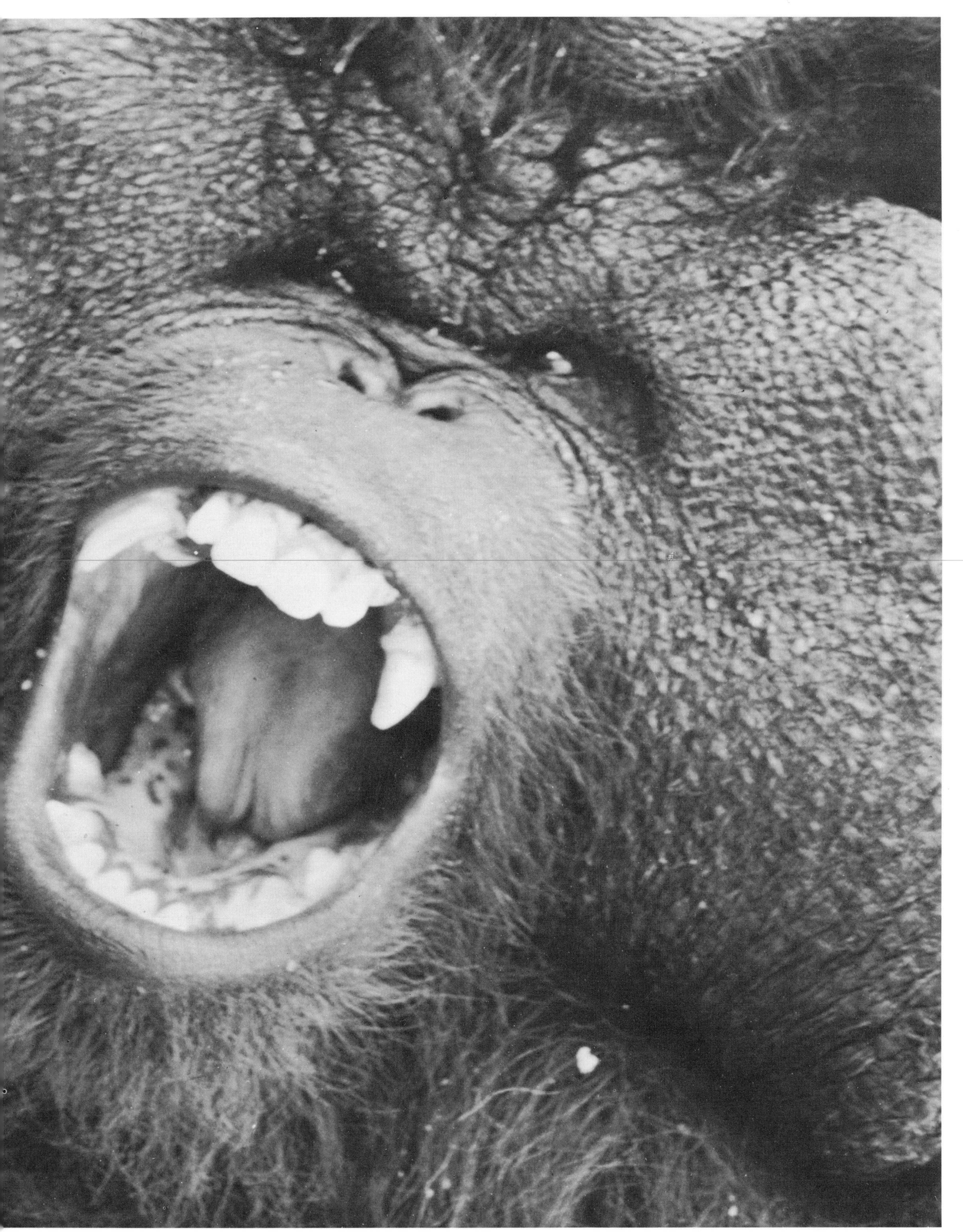

A warning snarl from a dominant male orang reveals the conical pointed canines that are used to rip open fruit.

Mias, he seizes it with his hands, and then bites it, and soon kills it.' Such beliefs may have derived from individual incidents, but they are not borne out by independent observation.

The general fear of the orang's great strength was rather more soundly based in experiences, for confrontations between man and orang were frequent. Wallace gained many specimens from natives who had attacked and killed lonely old individuals passing their remaining days closer to the ground. On one occasion a Dyak was nearly killed when he and his companions stumbled across an orang feeding quietly by a river. As the ape shuffled away from them, the group apparently closed in with spears, only to have the Dyak's arm severely lacerated by the orang's teeth and hand. His life was saved only by a timely assault by the others, who soon hacked the 'fearful beast' to death.

Wallace may seem to have been trigger-happy. In fact, compared to some of his contemporaries he was no such thing. During the greater part of the 19th century, scores of amateur hunters destroyed orangs purely for sport.

One was Captain Rodney Mundy who worked as the editor of Rajah Brooke's journals, but spent his leisure hours in pursuit of orangs. He tells of shooting females who still clutched their bullet-ridden young to their breasts, and of offering money rewards to anyone who could bring him a man-sized male, for he himself seemed to be able to find only females – all of whom he apparently slaughtered. Another was Captain Henry Keppel, also connected with Brooke. He talked of the immense size of the 'hand' in his possession and apparently used the uncertainties of ape classification – 'the facts are wanting and these facts I doubt not I can soon procure' – as an excuse for blasting away. In 1878 the American collector William Hornaday removed a total of nearly 50 orangs, dead and alive, from Sarawak.

Inevitably the orang population declined, the once fairly continuous distribution being cut up into isolated segments. No longer could their seasonal arrival at certain well-fruited areas be predicted, such as at the Banting Mission on the Lingga River, and no longer could the casual visitor expect to catch sight of them.

The 20th century brought a change in attitude toward the orang. Museums became saturated, and indiscriminate killing fell from favor. The demands for live orangs increased. But orangs are very difficult creatures to keep in captivity away from their native forests. The creatures died, and demand remained high. Prices spiralled. The drain of orangs continued. The combination of this drain of orangs and, more recently, severe habitat destruction, led the red ape to the very brink of extinction.

By the 1960's nobody had undertaken a serious ecological and behavioral study of the orang's life history, although such residents of Sarawak as Barbara and Tom Harrisson, who ran the Sarawak Museum, were fully aware of the need for one. In themselves, these two people contributed enormously, for they rescued orphaned juveniles, raised them with affection and sent them off to zoos where a life behind bars was infinitely preferable to a premature, mother-deprived death in the jungle. More recently the Harrissons began the idea of returning the matured orphans to the wild, there hopefully to live and to breed and maintain the population as best as possible. (This scheme, initiated by Barbara Harrisson in 1956 when she received her first orphan, is so vital to orang well-being that I mention it only briefly here. I will be enlarging upon it in Chapter 6.)

A pioneering study of orangs was not undertaken until well after such

THE LONELINESS OF THE LONG-DISTANCE CLIMBER

John MacKinnon, in 'In Search of the Red Ape', describes his experiences tracking lone orang-utans and discovering the details of their daily life as they make their way through the forests of Borneo.

'The big male proved difficult to find. I heard him calling several times during the next few days. He was slowly making his way up Tinggi stream, but whenever I took a compass-bearing on his calls and headed after him, he eluded me. It was four days before I finally tracked him down to a large fruit tree. Harold saw me at once but showed no fear. He was busily feeding on a type of wild lychee called *mata kuching* (cat's eyes). The ground below him was littered with several hundred discarded shells and stones, so he had obviously been there for some time. I noticed that he held his fingers in a curious manner but it was only when he at last climbed out of the tree that I realised they were completely stiff. The index finger of his left hand and two fingers of his right stuck straight out when the others were flexed, but this did not seem to impair his movement.

When I saw he was travelling directly towards me I became rather anxious. A thick liana passed over my head, bridging the gap between two adjacent trees and along this tightrope he now came. Swinging arm over arm he moved powerfully forward in a way that any gymnast would envy. It was too late to get out of his way so I crouched low and stayed quite still. His hanging legs can have been no more than a yard above my head as he passed, but he disdained to notice and carried straight on beyond me. For the remainder of the evening he fed in another *mata kuching* then sat resting quietly up the hill. There was a loud crash as a tree toppled over in the distance. Indignant, Harold raised himself high, facing the noise, and ballooned out his neck pouch with a loud, deep bubbling sound. Gradually this built up into the tremendous roars that I had heard so often. He reached a bellowing climax then gradually the groans subsided to low mumblings again. It was a great thrill to watch this magnificent display but I had to leave him to find a hideout for the night.

I climbed up the hill towards Horseshoe Ridge. The only level ground was a bare patch cleared by an argus pheasant as a dancing ring, and here I bedded down. In the valley Harold gave another long call while I covered my face with repellent to keep the mosquitoes at bay.

I was awakened by heavy crashing sounds from down by the stream. Only elephants could make such a noise and I suddenly felt very vulnerable. I was right in the open on an obvious pathway for any animal that should come up the hill. Hurriedly I collected my kit and, with the aid of my small pen-torch, moved off in search of a safer sleeping site. The side of the hill seemed far too steep for an elephant to climb so I settled there, by a small tree. Since I was in danger of rolling down the sharp slope in my sleep I arranged myself with one leg either side of the tree and lay down with my head pointing uphill. Almost asleep I felt a nudge at my foot. I kicked out to strike a frighteningly large resistance. An animal gave a tremendous snort and shuffled off with an incredible rattling noise. I groped frantically for my torch and in its dim beam saw a large porcupine snorting noisily only a few yards away. Bristling angrily, the beast plodded towards me. He was obviously not going to detour from his traditional path just because some fool was lying in the way. Unwilling to argue the point further I jumped aside and he trundled past me.

Once more I tried to sleep but the steep hill where I thought no animal could possibly balance was fast turning into a major thoroughfare. A tremendous crash below announced that this was exactly where my elephant wanted to come. Hastily I grabbed my belongings and moved on again. I paused at the pheasant ring where I had been earlier but not for long, as the elephant was still advancing steadily. My strange smell on the night air had aroused his curiosity; he was following at a slow but relentless pace. Panicking, I rushed off into the night, heading for the only safe place that I knew, the buttressed eyrie where I had spent my first night in the forest. I was afraid to use my torch with the elephant so close but fortunately knew the path quite well, or so I thought until I became caught up in a mass of *rotan* tendrils. This was no time for careful extraction so, with a quick flash of the torch to see my way, I lunged forward. I painfully tore myself free and hurried on across the narrow ridge and up the hill to my sanctuary.

At last I found the tree, clambered round the buttress roots and collapsed with relief on my bed of leaves. I was safe at last for, even if the elephant should follow me here, a strong liana provided an easy escape ladder up the tree. Thankfully I heard the beast crash past some way down the hill but, nevertheless, I got little further sleep that night. When morning broke I set off straight for home. There in the pheasant ring was a fresh pile of large droppings. Imagining elephants behind every tree, I broke all records for the distance and it was several days before I dared sleep out in the forest again. . . .

The fruit on the *bubok* tree behind my house began to ripen and it was this sweet bait that attracted Desmond, the only orang sufficiently intrepid to venture into our camp. Desmond was a fine young male with long, chestnut hair and he broached the tree late one evening. It was long after dark when he finished his dinner and settled down for the night in a large nest. This seemed an ideal opportunity for some photography, but before first light Desmond slipped away from camp and returned whence he had come. His next foray on to our territory resulted in a spectacular confrontation. Returning to camp I heard a terrific uproar of barking dogs and squealing orang. I rushed up to find Desmond trapped in an isolated tree, surrounded by three yapping dogs. Angrily he climbed down towards the excited animals and I hate to think what would have happened if I had left them to battle it out. Not wishing to witness a fracas, I pelted the dogs with bits of wood and as they scattered Desmond seized the opportunity and hurried off across the clearing to the safety of the forest. It says a lot for *bubok* fruit that, despite his experience, Desmond was back in camp feeding happily in the fruit tree a few hours later. This time he demolished all the remaining titbits, thus removing the need for another dangerous visit.

Like Desmond other large males frequently had to come to the ground for travel, but the females and young orangs were able to live virtually independently of the forest floor. These lighter animals did venture groundwards occasionally, however, to obtain small quantities of mineral rich soil. At the north end of my research area stood a limestone block sheltering a pile of hard, reddish earth. Imprints in this soil were unmistakably the toothmarks of orang-utans. Every month or so a new set would appear, evidence that another orang had visited the lick. Analysis of the soil showed it to be rich in sodium and potassium, which are rare elements in the acid forest humus and presumably of great value to the large mammals. The fact that orangs regularly visited such a local site indicated they had an excellent knowledge of the area's geography.

By the beginning of December I knew of several orangs now resident within my research area. Although the study was going well, fruit had become scarce and the weather was getting worse. There are few things more frightening than a thunderstorm in the forest. Violent winds toss the trees back and forth without mercy. Branches tear free and crash to the ground with terrifying frequency. Sometimes whole trees topple noisily to the floor, dragging smaller companions with them in a suicidal plunge, ripping an untidy hole in the forest canopy and completely blocking the ground passage. I always regarded falling timber as the most serious danger in the forest and several times had narrow escapes from falling missiles. Once while we were on the river a tall tree simply collapsed off the bank into the water ahead of us, swamping the boat before we had time to change direction.

The male orang-utans shared my dislike of falling timber and showed their displeasure with loud roars whenever they were disturbed in this fashion. When all other means of finding orang-utans failed I would simply sit in the forest waiting for a treefall and the subsequent protest to tell me whether I was in business or not.

It is the big old males who go in for such vocalisations. These high-ranking animals are up to twice the size of an adult female and their long hair, beards, high crowns, enormous inflatable throat pouches and enlarged fatty cheek flanges all attest to their status. Wherever they wander they announce their presence vociferously, warning other males to keep clear. Rivals may challenge the caller by bellowing in reply and at times the forest peace is shattered by up to three contestants shouting long-range threats at one another.

On the north side of the Segama River lived one particularly rowdy trio, Harold, Redbeard and Raymond. I met Harold up near the Tooth-Rock and he looked just as he had done the previous year, his fingers still stiff and his back still bare. I saw quite a lot of him in the next few months and as he was not afraid of me he was always a pleasure to watch. Redbeard was a more difficult customer and my first encounters with him were rather frightening. He was an incorrigible ground walker and as soon as he saw me would rush straight at me. Three times he chased me and three times I fled before I was able to repeat my tactics against King Louis and managed to bluff Redbeard into showing more respect. Between them these two split the region. Harold ranged from the centre ridge to the west while Redbeard's domain lay to the east. The single narrow ridge which served as the boundary was the scene of tremendous displays, Harold directing a barrage of calls east and Redbeard yelling westwards but they both took good care never to be there at the same time.

Raymond was a much quieter orang though very large. He did not visit the area often and showed no respect whatsoever for the Harold-Redbeard boundary ranging happily on either side and showing no fear of either of these heroes. This was rather amusing since he was quite terrified of me and would hide motionless at the top of a tall tree whenever I approached. One week when Harold was in residence on Central Ridge and had been calling several times a day, we heard the unmistakable tones of Raymond nearby. His solitary bellow had a dramatic effect for it was ten days before Harold dared give tongue again. Redbeard was no braver. When he heard Raymond calling in the distance he chorused a reply and hurried to investigate this rude intrusion. As Redbeard drew closer, Raymond called again. This time Redbeard must have recognised who it was for he turned tail and fled back the way he had come.

Obviously strong jealousies and competition existed between the males. As a spacing mechanism their calling behavior was very effective for adult males practically never met each other. On the rare occasions when they did the encounter was always accompanied by terrific branch-shaking displays until one of the antagonists turned and fled. Probably the great size, long hair, enlarged faces and ugly expressions of male orang-utans are designed to strike fear into the hearts of their opponents. However, Nature does seem to have over-excelled herself as the female orang-utans are also frightened by their fearsome brothers and it is the younger sub-adult males who have more success with the ladies. Adult females sometimes fed near to Redbeard but Raymond and Harold were always alone. Nor did Harold desire such frivolous company, for on one occasion when he heard a female and juvenile moving towards him he climbed smartly to the ground and slipped silently off before they saw him.

I came to know some of the orangs, like Harold, quite well and learned to recognize their particular habits and idiosyncrasies. But many of the animals were temporary visitors whom I would see for only two or three days before they left my part of the woods for good. Meanwhile I had established my own range with my own system of paths and my own favoured haunts. There were places where I habitually drank from streams or cut refreshing vines for their sweet sap. I had special lookout points and listening spots, favorite log benches and tree blinds. With the help of Bahat and Pingas I erected several polythene-covered shelters at convenient sites throughout the region. Here I stored caches of tins so that food and shelter were always at hand wherever I might be working. . . . Suddenly in June the population became more active. The males began calling vigorously and many unfamiliar orangs arrived from the west. Sarah, the lone female, was the first to meet the invaders. She squeaked excitedly and shook branches at a strange female and juvenile passing through the trees below her. Half an hour later another female with two young appeared, closely followed by two large males displaying and branch-shaking as they progressed. The red sub-adult was obviously fed up at being pursued by his angry elder and took out his ill-feeling on Sarah, who scurried away protesting at the injustice. The big, black male decided that

An orang's arm appears above its nest.

I would make a better victim. There were eight orangs in the neighbourhood, a wonderful opportunity for observation, but when the menacing ape climbed to the ground to meet me I decided it was time to leave.

Within a mile of this unfriendly crowd I spotted a familiar brown shape clinging to the white bark of a *Polyalthia* trunk. Midge was a few feet below his mother and both were happily engaged in chewing strips of bark, sucking out the sweet sap and spitting out the remaining wodges, just like old men chewing on their tobacco. They seemed perfectly at ease but cannot have remained unaware of the presence of other orangs for very long for the next three days were frequently disturbed by the booming calls of the aggressive black male. Margaret and Midge meandered slowly towards their old haunts on Horseshoe Ridge. Eventually I became bored with their shyness and sluggish pace so headed north again to see what the newcomers were up to. I had another amazingly full day for they were still moving in a tight-knit group and I was able to watch nine different animals. Most were timid females accompanied by their young but a sub-adult male, who I felt must be Humphrey of the winter before, was very tame and let me sit close to him while he enjoyed a meal.

Day after day I followed different members of this band, which was rolling slowly eastwards. Individuals wandered wide to north and south but there seemed to be a definite group nucleus where perhaps half the emigrants would be congregated at any one time. This was quite different from anything I had seen before in an entire year's fieldwork. I caught one more glimpse of my black antagonist storming past along the ground, too intent on his mission to notice either me or the two orang-utans swinging above. After that I saw no more of him but the other travellers continued steadily and crossed Central Ridge into Redbeard's domain. Redbeard seemed to have been expecting them for I had heard him calling several times during the preceding week. He certainly gave them quite a welcome, hurtling through the trees to attack two helpless females and their young. They fled screaming up a narrow valley but Redbeard chased after them remorselessly. Catching up with the smaller female he dragged the unfortunate miscreant out of the tree and beat and raped her before moving on up the ridge, calling as he went.

Ironically in this time of famine for the bearded pigs, the sweet fruits they would have welcomed were gorged on by orang-utans and monkeys before they had a chance to fall. There were even few of the less popular, oily dipterocarp seeds which maintained the pigs at other times of the year.

The climate in tropical rainforest remains very similar throughout the year. Although in the dry season there is only half as much rain as during the wetter months, the dark jungle is always moist and the temperature stays much the same, rising to about 90°F in the heat of midday. It is so damp that leather quickly perishes and camera lenses support flourishing growths of fungi. Nevertheless, in spite of the year-round uniformity of heat and humidity, definite seasons do prevail and are no less obvious than those of the temperate regions.

Many of the trees have their own appointed times for flowering, fruiting and putting out new leaves so that the commoner species can greatly affect the flavour of the whole forest. When the *Melapi* trees are in bloom whole hillsides are white with their blossom and the *Ramus* vines decked in new leaves tinge the forest with red. Since most seeds ripen between April and November there is a recognisable fruit season comparable to a temperate summer. This time of plenty is heralded by the *Wadan*, the tall, climbing bamboos, whose hard, nutty produce are a common and important item in the diet of the orangs. As the months pass the wild plums, lychees, *rambutans*, taraps, *lansats*, figs and durians all ripen to yield a generous harvest. But with the approach of another rainy season the wealth and variety of fruits starts to drop off so that the animals must make the most of the lavish crop while it lasts.'

work had been done on gorillas and chimpanzees, for there were formidable obstacles to be surmounted. Firstly, there was so much support for primate study in Africa that the Far East was rather neglected. Secondly, the orang was always going to prove an awkward customer to study in the wild due to its solitary nature, sparse distribution and unpredictable passage through the treetops – all of which combine to make it a completely different proposition from its African relatives, with their tighter social organization, more terrestrial habits and smaller natural ranges.

The man who emerged to undertake the task was a young Oxford graduate, John MacKinnon, grandson of a British Prime Minister, James Ramsay MacDonald. MacKinnon's first trip to the Far East was as an undergraduate backed by the University Expeditionary Council. In mid-1968 he found himself alone on the banks of the confluence of the Bole and Segama Rivers in Sabah, North Borneo, with several days before his native guides could catch up with him in their slow canoe. Still pink from exposure to the tropical sun, and inexperienced in the ways of the jungle, he slowly adapted, paying the usual price for ignorance and uncertainty. He recalled later how leeches looped thirstily toward him wherever he went; how finely-barbed tendrils of the *rotan* palm clawed at him, holding him tighter and more painfully as he fought to free himself; how mosquitoes and sandflies fed on his body when in the evening he collapsed exhausted onto his make-shift camp bed.

He soon found his orangs basking lazily among the boughs of fruit-laden trees, suspicious, but neither frightened nor aggressive. He saw something of their way of life, attuning himself to their movements and their moods. By the time he returned to England after his first brief visit he had probably more hours of observation than anyone before him.

After his final examinations in zoology at Oxford University he immediately set off again for the East, where he planned a full-scale doctoral study of orangs. In the years that followed, during which time he says he often pondered whether he was becoming one of them himself, he patiently tracked, watched, mused and took painstaking notes, films and photographs until finally he had built a picture which today forms the focal point of all that is known about the red ape of Borneo and Sumatra.

It was the difficulty with which MacKinnon found his subjects that provided the very clue to their way of life. Once he had located them, he stayed in touch, even if to do so involved days of travelling, sleeping out and enduring maximum hardship with no more than minimum provisions.

If any one factor can be singled out as the key to orang behavior, it is the distribution and seasonal ripening of forest fruits. The apes show a strong liking for such exotics as fig, lichees, plum, mangosteen, durian and rambutan, all lone trees scattered somewhat thinly through the forest. To survive they must know exactly where these delicacies are located and when they will be available. MacKinnon is certain that orangs carry within their heads a distribution map of each species of tree. When they find one of them ready for eating (it need not necessarily be completely ripe – much to the anger of the more particular natives) they will immediately strip it bare and then set off for the remaining trees of the same species growing within their range.

MacKinnon once made a map of the highly prized durian trees in one location. When the fruit began to swell, a subadult he had nicknamed Humphrey, who had not been in the area for years, realized the fruit was ready and then led MacKinnon on a 'forest day' which took in no less than

A mother's protective arm symbolizes the care and devotion that goes into her seven-year task of raising a young orang.

six of the 18 trees available, heading almost directly from one to the other. The purpose of the trip was clear – the different fruit trees do not ripen at the same time. While the fig ripens every few months, other fruits may not do so for years at a time. The brainpower that can store such complex information so precisely and recall it instantly after many years must be considerable.

The slow speed with which the orangs travel through the trees is at least one clue to their solitary natures. A whole family unit would never find enough to eat in any one small area.

In Sumatra the orangs are less solitary. On a visit there in 1971 MacKinnon was struck by the much more closely-knit nature of their social life. He saw 14 different individuals feeding in one tree over a period of five days, a degree of crowding never seen in Borneo. He suggests as an explanation the increased dangers of predation from such animals as the tiger and increased competition from the lesser ape, the siamang, which he noticed was capable of putting a single orang to flight from a tree.

But it is obviously not enough to evolve a system which perfects just one vital aspect of a whole way of life. Within the solitary and nomadic feeding framework, orangs must breed and care for their young. Unless these operations can be carried out successfully, there can be no future at all. So, like all animals, orangs must find a compromise – one which permits the union, however fleeting, of male and female and which commits the female to raising the slow-maturing offspring.

A full grown male of advanced years – orangs may live until they are at least 40 – will pass his days wandering in a leisurely fashion through the treetops. Occasionally he will pause to give voice, perhaps a series of bellowing groans rising in tempo and trailing off in a succession of guttural cries. These vocal signals may carry for as much as a mile through the jungle, letting other orangs know where he is and even what his mood is.

It is easy to assume that the males are calling to attract females and to maintain an acceptable distance between themselves to avoid overcrowding. But MacKinnon soon realized that while the older males were engaging in their vocal warfare, it was the younger and less vociferous males who were actually consorting with most of the females. Clearly the females were more attracted by youthful silence than by aging clamor. As the biological success of an individual is usually measured in terms of the number of its young who themselves survive to reproduce, it is difficult to see what the older males get out of literally frightening females into the arms of younger suitors. MacKinnon suggests that by their behavior the older males, no longer able to compete on purely physical terms, attempt to disrupt the courting behavior of their neighbors in order to get the female, as it were, on the rebound. The result of this disruption is that an area of forest maintains a population lower than it would if the behavior did not exist, thereby giving the next generation more space in which to live and breed successfully.

This is an interesting hypothesis because, if true, it shows that the survival of the species depends on more than mere sexual success. The old male becomes the equivalent of a sage. The young that he has sired take many years to mature and during that time his stored knowledge of ripening fruits in his unchallenged domain plays a vital role in their welfare. At some stage, the best way to ensure the spread of his genes is not to father more children, but to safeguard the future of his already created offspring, even if that requires relinquishing the bulk of his breeding rights to someone else. Biologically success is a complex phenomenon and it cannot be judged

Responses to the Territorial Imperative

Biruté Galdikas-Brindamour, an anthropologist from the University of California, Los Angeles, who runs an orang rehabilitation center at Tanjing Puting, describes some confrontations:

'Perhaps my most vivid memory, though, is of that scorching day I came face-to-face with a large adult male on the ground. It was almost a showdown. I was rounding a turn in a ladang path when a huge orangutan appeared, heading straight toward me. He was just ambling along, head down, oblivious to my presence. Then he stopped dead in his tracks less than twelve feet away. For long seconds he stared and stared. . . . Strangely, I felt no fear. I simply marveled at how magnificent he looked with his coat blazing orange in the full sunlight.

Abruptly, he whirled around and was gone. There was nothing but the sound of his feet padding off along the path.

My confrontation with this big male seemed to bear out a traditional belief that the wild orangutan is mild and retiring. Back at camp, though, our workman, Ahmad, told us of a relative in Kumai who had lost half his hand and part of one foot to an orangutan male he had encountered in a ladang. But it turned out that the relative had been chasing the animal with dogs. A full accounting of incidents like these always led to the same conclusion: Humans who were bitten or wounded had invariably provoked the apes.

By now, meeting an orangutan on the ground came as no surprise. One mature male sometimes spent as long as six hours a day traveling and foraging on the ground, though on other days he stayed totally in the trees. Another large male came down from the trees daily and did almost all his long-distance traveling along the forest floor.

I was, however, amazed to see a subadult male sleep for 45 minutes on the ground during the day. He didn't make a nest but merely bent a sapling under him as he lay down. This was the first time that anyone had seen a wild orangutan sleeping on the ground. Since then we have found three actual "ground" nests. In these instances the nests were built on fallen logs less than a yard off the forest floor. . . .

Early every morning I went out in search of wild orangutans. As I walked I listened. The wild orangutans sometimes disclose their presence by the snap of a twig or the regular dropping of fruit stones as they eat, sometimes by the crashing of branches as they move through the trees.

Once I had located an orangutan, I followed it, if possible, until it nested for the night. Next morning I'd be back before sun-up, hoping to find the orangutan still in the nest, so that I could continue to follow and observe its behavior. I would sometimes walk three miles to reach a nest by the first glimmer of dawn.

On one such search I encountered a pregnant female and her juvenile son. We named them Cara and Carl and started a continuing study of them. Cara was a problem. She had a dangerous habit of breaking off branches and dropping them without any of the usual vocal warnings. One dead branch – a veritable log – missed me by about an inch.

Cara, unlike most females, also toppled branchless dead trees. Once, when I was tangled up in a windfall below her, she suddenly started rocking one such enormous snag in my direction. Had she the strength of the much heavier males, I would have been killed. As it was, the snag teetered but didn't fall.

Despite Cara's initial hostility, she and Carl after about a year and a half came to accept us more fully than did most of the others. And they gave us our first glimpses of social interaction among wild orangutans. One instance sticks in my mind. The two had been joined by a subadult male, one of several who occasionally followed Cara when she was in heat. The newcomer romped good-naturedly with Carl and feverishly examined Cara's genital region. At the end of the day the animals bedded down close together. Cara and Carl occupied one nest, the male another.

All was still. It was slowly becoming darker. Suddenly some trees began to shake as if in a hurricane. Snags crashed and there was a piercing bellow as an adult male emitted his "long call" – a hair-raising, minutes-long sequence of roars and groans that can carry a mile. The subadult male didn't hesitate; he leaped out of his nest and dashed down to a perch a yard above the ground. He listened until the call ended, then slid to the ground and vanished in the undergrowth.

Amid wild shaking of trees and snapping of branches, a gigantic male emerged. He had the pronounced cheek pads and throat pouch that males acquire when they reach adulthood at age 12 to 15; the throat pouch probably acts as a resonator for the long call. He peered over the edge of the occupied nest. Cara and Carl could not have slept through the commotion, yet their bed never once quivered or shook. Satisfied this was only a female and her offspring, and not a sexual rival, the big male moved away.'

in the light of one factor alone.

Whatever the reasons for this mating strategy, the Dyak natives have their own refreshing interpretation of the calls of the male orang-utan. According to their legend an old male once abducted a young girl from a village, keeping her alive in the forest on a diet of wild fruit. In return she produced for him a baby – half human, half orang. But one day she escaped from his devoted vigilance and ran toward the river with her enraged kidnapper in hot pursuit, natives cried out for her to abandon the child and to leap into their boat for safety. This she did, thereby making good her own escape. But her incensed 'husband' snatched up the child and in his passionate rage tore it in two, hurling the human half after his vanishing bride and the orang half back into the forest. To this very day, heartbroken, he roams the trees alone, calling pitifully for his departed love.

When it is first born, after a pregnancy of eight or nine months, a baby orang clings limpet-like to its mother. She continues her treetop wanderings as though unburdened by the 2 to 4 pound infant attached to her side, its small face peering passively at the jungle around it. When she stops to feed or to rest, its natural urge to explore and to strengthen its limbs takes over and it clambers around her hairy body. Sometimes, as it gets older, it samples her meal of fruit, but during at least the first year of its life it relies almost totally upon a supply of milk from her breasts for nourishment.

In the first months of its life the youngster increases in strength and in awareness of its surroundings. The end of its first year brings an exploratory phase of childhood. The physical dependence on its mother is relaxed as

it ventures with unsteady limbs into the forest canopy, hesitating and testing its way with increasing confidence. The growing urge for playmates is satisfied only by the presence of its mother, unless an elder brother or sister happens to be around and willing to tolerate juvenile frolics.

During its second year, the youngster is less dependent upon its mother's milk and able to feed on some of the jungle fruits around it. Perhaps the best game of all is breaking branches inward to form the first nest that it will build almost every night when it is alone in the forest. During early days, the youngster watches its mother perform the brief but effective operation and as it is older and stronger, it learns the same technique – the key to its future independence. The first attempts are infused with playfulness as branches are bent haphazardly, without serious attention to economy and mechanical precision. When finally the crude semblance of a temporary home has been constructed, it may well be abandoned for the warmth and security of that made by its experienced mother close by. (Some of these nests may be 60 feet above ground, hurriedly pieced together at the end of each day's travelling. In rainy weather orangs may cover themselves with protective leaves, for which reason the natives say they build houses in the trees.)

As the youngster grows into its third year, the ties with its mother are reduced dramatically. She may well have conceived again and, if so, is now more concerned with the impending responsibilities of motherhood than with the previous offspring. The youngster, of course, objects to this rejection and continues to engage her in play, even attempting to suckle and stay

Two shots show the success of breeding programs in zoos. At left seven-month old twins, strong and healthy, dangle from a bamboo pole in Munich's Hellabrunn Zoo. At right, a baby born in Frankfurt Zoo is held by its new foster mother after its real mother proved unwilling to feed it.

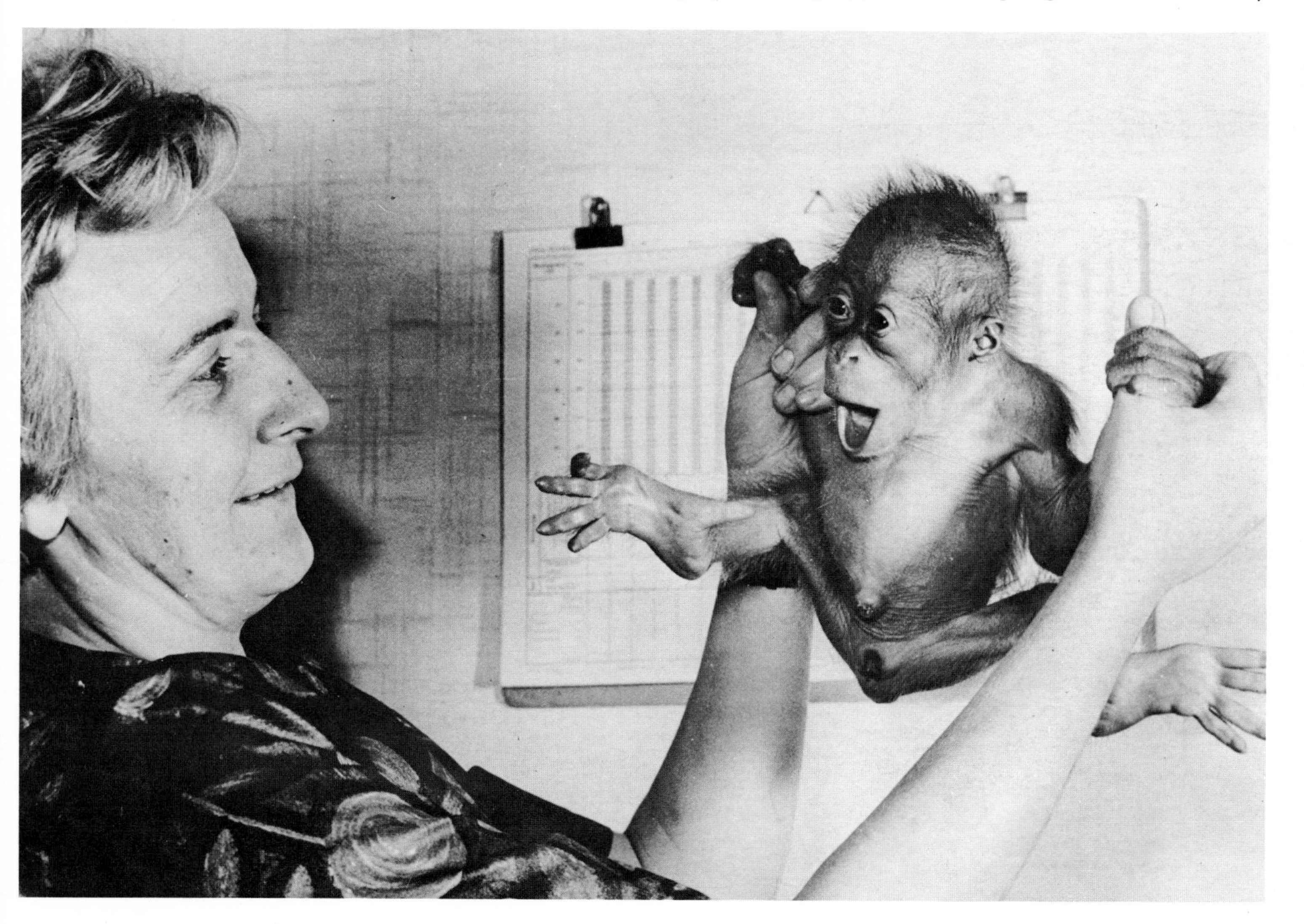

There are now several hundred orangs in zoos around the world, but the lives of such creatures – like the ones in these two pictures – bear little relation to their lives in the wild. Some zoologists suggest that the only hope of saving orangs is to restock wild populations with zoo-bred animals.

close to her. But the arrival of the new baby signals the end of the close, protective relationship.

A young male will now be left to his own devices, forced to travel further afield, finding his own food and adapting himself to his solitary future. A young female, however, will not stray so far and will accompany her mother, often playing with the new infant and learning the way it should be treated. This is undoubtedly a vital time for orang welfare, for when this inexperienced female, at about the age of seven, comes to give birth herself, there will be nobody around to give her any advice.

The young male continues to grow well beyond the age of ten. Then the hallmarks so characteristic of his sex appear – the shaggy coat, the bulbous face swellings and the strengthening voice. He has already moved up in the hierarchy and the search for willing females, which once found, may take considerable persuading, is on. So, too, is the competition with other aspiring males eager to stake their own claims in the forest.

One wonders how the orang-utans can possibly survive now that their distribution has become so fragmented. Perhaps the days are nearly over when proud old males call through acres of untouched forest.

CHEEKY SIGNS OF MALE POWER

One of the most extraordinary characteristics of mature male orangs is the way their faces acquire pads of flesh, dewlapped double chins and cheek pads which in extreme cases form massive, fatty 'blinkers' that almost obscure the diminutive eyes.

There are good reasons for the development of such characteristics. Threat displays between males involve visual contact and the more exaggerated the ornaments, the more awe-inspiring they will appear to rivals. Females are more likely to yield to the biggest, most awe-inspiring males.

The blinkers are deposits of fat which are not only important for display purposes, but also contain valuable reserves which may be drawn upon during lean times of fruit supply. But in zoos, when the males are inactive and continuously well fed, the fatty pouches can overdevelop until they dangle like the jowls of obese old men. Their significance lost, they become only a hindrance to their owners and objects of amazement to onlookers.

A zoo-reared orang displays the exaggerated cheek pads that result from many years of good eating.

Beneath the cheek pads, male orangs also have a pendulous dewlap – a laryngial airsack which may be used as a resonance chamber when calling. Much larger in the male

than in the female it can also develop into a fatty excrescence that hangs like an apron down over the chest of the male.

3/GORILLAS: THE FOREST GIANTS

The gorilla, largest of all the primates, is an awesome physical specimen. A full-grown male (left) may weigh as much as 600 pounds. His head adds to his menacing look, with its high forehead and small ears set flat against the sides. Yet despite their great size and fierce aspect, gorillas have remarkably peaceful dispositions and lead generally tranquil lives. Found only in the tropical forests and mountains of central Africa (below), they are strict vegetarians. Because of their size, they feed and live mainly on the ground in troops of five to 30 individuals, eating chiefly tree ferns and vines.

Anyone who set foot in 18th-century Equatorial West Africa heard of gorillas. Travellers were regaled with horrendous accounts of the physical strength and sexual prowess of the 'Wild Men' of the coastal forests. The natives knew these 'Wild Men' as *Ingena*, their word for almost everything that is wrong in a civilized human being.

Today the word 'gorilla' conveys a rather different image. The gorilla is the largest and most powerful of the anthropoid apes; it is a shy and retiring primate, with a none-too-certain future, confined to the forested regions of western and central Africa. But a great deal of drama, fear, confusion and controversy have accompanied the process of scientific identification over the last two centuries.

The first stories referring to such a creature by name found their way out of Africa 2500 years ago, when the first tentative steps were being taken to explore the globe. In the 5th century BC Hanno, a Carthaginian explorer and colonizer, was on a voyage around Africa when he called in on the Equatorial West coast, probably near the Gabon River. He wrote: 'We came to a bay called the Horn of the South. In the recess there was an island . . . having a lake, and in this there was another island full of wildmen. But much the greater part of them were women with hairy bodies, whom the interpreters called Gorillas. But, pursuing them, we were not able to take the men; they all escaped, being able to climb the precipices, and defended themselves with pieces of rock. But these women (female Gorillas) who bit and scratched those who led them, were not willing to follow. However, having killed them, we flayed them, and conveyed the skins to Carthage.'

This drawing, done in the second half of the 19th century, gives us a surprisingly accurate idea of the physical structure of a female gorilla, without dramatizing the more superficially frightening characteristics of the creature.

By the very nature of their behavior, Hanno's 'gorillas' were probably none other than baboons. There is some speculation on this point. The German zoologist, Bernard Grzimek, suggests that they could not have been baboons because Hanno and his men would have been familiar with these primates as a result of extensive trading in North Africa. Nonetheless the message was clear: Africa harbored human-like creatures which struck fear into the hearts of even the natives.

The image of Africa as a continent of monsters lasted until well into the 18th century, for most authorities reproduced parrot-fashion the opinions of classical authors. According to Pliny, the Roman naturalist, African fauna included *pegasi* – winged horses with horns – and the terrifying *mantichora* which 'has a triple row of teeth, meeting like the teeth of a comb, the face and ears of a human being, gray eyes, a blood-red color, a lion's body, inflicting stings with its tail in the manner of a scorpion, with a voice like the sound of a pan-pipe blended with a trumpet, of great speed, with a special appetite for human flesh.' Solinus, the 3rd-century Roman author, mentions in his book on natural history 'ants as big as a mastiff that have talents (talons) like Lyons.' The native men of Africa were believed to be equally strange. Pliny records the presence of *Blemmyes* – men who 'by report have no head but mouth and eies in their breasts.' And finally a certain de Monas appended a note to his 1761 map of Africa about the evolution of monsters there: 'Scarcity of water forces different animals to come together to the same place to drink. It happens that finding themselves together at a time when they are on heat, they have intercourse, one with another, without paying regard to the difference between the species. Thus are produced those monsters which are to be found there in greater numbers than in any other part of the world.' Africa was so riddled with deep mysteries that anything, even giant men of untold strength and vile habits, was possible.

The myths still live on, for they have superficially more appeal than the truth. *King Kong*, made in 1933, established the gorilla firmly in the popular consciousness as an avenging monster (although interestingly, King Kong came not from Africa, but from a Southeast Asian island). The film's remake in 1976 has no doubt confirmed that awesome reputation in countless minds.

Truth only began to separate from fiction around the beginning of the 17th century, when an English sailor named Battel (though the exact date is uncertain and the name sometimes corrupted as Bartlett) was captured by Portuguese in Angola and reported native stories which seemed fully to confirm the more northerly existence of strange forest beasts. He reported two monsters of which the natives were terrified. One was the chimpanzee; the other was the *pongo*, probably the gorilla, which '. . . is in all proportion like a man, but that he is more like a giant in stature than a man; for he is very tall and hath a man's face, hollow eyed, with long haire upon his brows. . . . They sleep in the trees and build shelter for the raine. They feed upon the fruit that they find in the woods, and upon nuts, for they eat no kind of flesh. They cannot speak and have no understanding more than a beast. The people of the countrie, when they travaile in the woods, make fires when they sleepe in the night: and in the morning when they are gone, Pongo will come and sit about the fire till it goeth out, for they have no understanding to lay the wood together. They goe many together and kill many negroes that travaile in the woods. . . . These Pongos are never taken alive, because they are so strong ten men cannot hold one of them; but they take many of their young ones with poisoned arrows. The young Pongo hangeth on its mother's belly with its hands clasped about her, so that when any of the country people kill the females, they take the young which hangs fast upon his mother.'

This account seems to refer to the gorilla, for it is in many ways highly accurate. Battel may even have combined native stories (of ferocity) with his own personal observations (of feeding and nesting).

A century later Africa was becoming more accessible to Europeans and by the early 19th century, explorers were penetrating the interior of the continent. But the truth about gorillas did not surface immediately. In 1819 a certain Bowditch journeyed from Cape Coast (in present-day Ghana) inland to Ashanti territory and heard nothing but myths about gorillas: of how they lurked in the undergrowth to destroy travellers, not for food, but – one must suppose – just for fun. His natives swore that all forest killings were attributable, not to inter-tribal rivalries, but to the great ape, for, 'Sometimes . . . when a company of villagers are moving rapidly through the shades of the forest, they become aware of the presence of the formidable Ape by the sudden disappearance of one of their companions, who is hoisted up into a tree, uttering, perhaps, only a short choking sob. In a few minutes he falls to the ground a strangled corpse, for the animal, watching his opportunity, has let down his huge hind-hand and seized the passing negro by the neck with a vice-like grip, and has drawn him up into the branches, dropping him when life and struggling have ceased.'

In the early decades of the 19th century there were no reliable reports of a white man having actually seen a gorilla, either dead or alive. Coastal missionaries believed – upon native 'evidence' – that gorillas were alienated human beings, cast out by members of their own race to lead degenerate lives in the forest, there either to roam in murdering bands, exacting their revenge upon the people who had turned them away from the civilized world, or to skulk in the most inaccessible places, ashamed to come out into the open for fear that they might be recognized.

We must be grateful to an unnamed American missionary for the first permissible evidence of the gorilla's existence. He was one of those dedicated followers of Christ who pursued an interest in natural history beyond the normal call of duty. He had certainly heard all about gorillas for, in 1847, when a large skull came into his possession '. . . it struck him as being so extraordinary that he believed the natives were correct in attributing it to the much-talked of Ape of whose ferocity and strength he had heard so much.'

Gorilla country in Equatorial Africa includes the Kahuzi-Biega National Park in Zaire (left) with its steep hillsides clothed in lush forest. Here, and in nearby areas, gorillas can lead a leisurely existence, often almost inconspicuous (right) as they feed on the rainforest floor.

The specimen found its way into the competent hands of one Dr Savage, an amateur anatomist who had lived for many years in the vicinity of the Gabon River. He too knew all about gorillas and yet the skull that he now held in his hands was so reminiscent of that of the orang-utan from Southeast Asia that for a while he labored under the mistaken impression that they were one and the same animal. As he became more acquainted with the intricacies of skull proportion and assembly, he realized how much closer it was in shape, if not in size, to a chimpanzee skull. Upon his return to America, Savage began extensive comparisons between the two animals and entered into correspondence with Richard Owen, England's leading anatomist of the day, who had also received some gorilla specimens from the same part of Africa. Between them, and apparently in complete agreement,

they decided that the gorilla was sufficiently distinct from the chimpanzee to warrant naming it as a separate species and yet sufficiently similar to the chimp to be retained within the same genus. The chimpanzee had already been described as *Troglodytes niger;* the gorilla was now called *Troglodytes gorilla*. The choice of the name *Troglodytes* (later changed) was a rather strange one. A troglodyte is a cave-dweller; at the time the name was applied to an Ethiopian tribe whose claim to fame was that it lived in mountain caves.

After centuries of fantasy, the reputation of the gorilla was gradually being brought under the scrutiny of scientists. Their responsibility was to view the facts with an objective eye. They did not have much to go on – just one skull and a few bones. More hard evidence was needed.

In 1855 Paul du Chaillu left America for western Africa. He was probably more of a traveller than a scientist (and he was later to be proved something of a charlatan) but he was, nevertheless, a brave and determined man. His expedition had the distinction of being the first to arrive in the tropical zone with the sole intent of exploring gorilla country and of obtaining the great ape either dead or alive.

At the headwaters of the Ntambounay River, du Chaillu came across an old and deserted village. Its sugar-cane field lay in a disorderly fashion. Something about this field attracted his attention. It had evidently been

Paul du Chaillu, the French 19th-century explorer, devoted much of his life to hunting and observing gorillas and chimpanzees.

Trapped uncharacteristically up a tree, a lowland gorilla makes an easy target for a great white hunter, accompanied by a bevy of rather apprehensive native guides.

abandoned some time before because it was now being clawed back by the encroaching forest. But in places there were signs of some sort of activity Was it of humans or of non-humans? Du Chaillu searched around for evidence. There, before him, canes had been pulled up and strewn about. Their ends had been chewed; incredibly, they were fresh. His intuition told him that he had almost stumbled across a feeding party of gorillas and that they, more sensitive to his approach than he to their existence, had slipped quietly into the sinister gloom of the jungle. It may have occurred to him, briefly, that the gorillas had taken over the village for themselves and even now might be watching him and his small band of companions.

But he turned his attention towards the field. 'I knew that they were fresh tracks of the Gorilla,' he wrote later, 'and joy filled my heart; they (the native hunters) now looked at each other in silence and muttered, *Nguyla*, which is as much as to say in Nepongwe, *Ngina*, or as we say, Gorilla. We followed these traces, and presently came to the footprints of the so-long desired animal. It was the first time I had ever seen these footprints, and my sensations were indescribable. Here was I now, it seemed, on the point of meeting face to face that monster of whose ferocity, strength and cunning, the natives had told me so much; an animal scarce known to the civilized world, and which no white man before had hunted. My heart beat till I feared its loud pulsations would alarm the Gorilla and

Congolese tribesmen display a lowland gorilla in their village in about 1910, one of the many trophies taken by big-game hunters in the early years of the century.

my feelings were excited to a painful degree. By the tracks it was easy to know that there must have been several Gorillas in company. We prepared at once to follow them. The women were terrified, poor things, and we left them a good escort of two or three men to take care of them, and reassure them. Then the rest of us looked more carefully at our guns, for the Gorilla gives you no time to re-load and woe to him whom he attacks. We were armed to the teeth . . . I knew that we were about to pit ourselves against an animal which even the leopard of these mountains fears, and which perhaps has driven the lion out of his territory; for the kind of beasts so numerous elsewhere in Africa is never met in the land of the Gorilla.'

Picking his way cautiously through the bush, along a stream and over the brow of a densely vegetated hill, du Chaillu entered a world of huge granite blocks where he supposed the gorillas must be hiding, perhaps even waiting to ambush and destroy his persistent little band of men. Suddenly the air was punctuated by a 'strange discordant, half-human half-devilish cry' and four gorillas broke cover and made for the safety of the dense forest. Du Chaillu fired blindly and inaccurately. He set off in hot pursuit, only to lose his quarry in the woods.

Poor du Chaillu! He had come so close to his objective but now, having no doubt incurred the gorilla's wrath, he was forced to spend the night in a make-shift forest camp. He and his quaking companions huddled round a roaring fire which they fed eagerly with anything that would burn. They went again and again over the events of the day until finally they lost touch with reality. The lone white man became the object of all the natives' fears as they poured forth tale upon tale of gorilla atrocities, feats of strength and elusiveness.

It was enough, one would have thought, to deter the average hunter from continuing his jungle folly thousands of miles from home. But du Chaillu was no average hunter. He listened patiently, albeit nervously, perhaps even slightly amused that a mere animal could reduce grown men to such a state of submissive inferiority. As the black of night gave way to the first protective wash of dawn, he reasoned that the greater part of the myths surrounding the gorilla originated in the overimaginative, superstitious and none-too-courageous local blacks. He stretched his aching limbs, more determined than ever to meet the gorilla face to face.

It was another week or so before du Chaillu found himself again following fresh gorilla tracks through the forest. This time his quarry appeared to be a large, single animal and the now more experienced hunter proceeded with caution. He and his party crept quietly through the undergrowth. Suddenly the air all around them was filled with a furious barking and roaring. The bushes immediately ahead swayed violently and amid a crashing and tearing of branches, the lone male towered in front of them.

'It stood about a dozen yards from us, and was a sight I shall never forget. Nearly six feet high, with immense body, huge chest, and great muscular arms, with fiercely glaring large deep gray eyes, and a hellish expression of face, which seemed to me like some nightmare vision; there stood before us the king of the African forest. He was not afraid of us. He stood there and beat his breast with his huge fists till it resounded like an immense bass drum, which is their mode of offering defiance; sometimes giving vent to roar after roar. . . . His eyes began to flash fiercely for we stood motionless on the defensive, and the crest of short hair which stands on his forehead began to twitch rapidly up and down, while his powerful fangs were shown as he again sent forth a tremendous roar. He advanced a few steps, then

Fanciful drawings like these two built up an image in the public mind of the gorilla as a ferocious and vengeful monster. At right, a gorilla savages a black hunter and, at far right, a Tarzan comic-strip gorilla – 'Kerchak, with an iron hand and bared fangs' – goes on the rampage.

stopped to utter that hideous roar again; advanced again, and finally stopped when at the distance of about six yards from us, and then, just as he began another of his roars, beating his breast with rage, we fired and killed him. With a groan which had something terribly human in it, and yet was full of brutishness, he fell forward on his face. The body shook convulsively for a few minutes, the limbs moved about in a struggling way, and then all was quiet; death had done its work, and I had leisure to examine the huge body. It proved to be five feet eight inches high, and the muscular development of the arms and breast showed the immense strength it had possessed.'

During the remainder of his time in the forests, du Chaillu secured specimens dead and alive. We owe him a great debt for providing so much first-hand information about the gorilla. He sent to England the prepared skins and bones of at least five specimens which became the focal point of much scientific controversy. There were those who revelled in every word that emanated from the great hunter's pen and there were those who still could not bring themselves to believe in the existence of such a beast.

But whatever the opinion of the armchair scientists, du Chaillu's reputation seemed secure. He alone had succeeded where for centuries men had failed before him. Revelling in his new-found limelight he revealed a major fault – he was overimaginative. This was to manifest itself in such a number of important ways that a shadow was cast over the credibility of all that he said and did.

On the limited experience of keeping a young gorilla in captivity – his 'pet' was actually three years old and not very co-operative when placed in a small cage – du Chaillu declared that gorillas could never be kept success-

In this apocryphal incident, retailed as true by du Chaillu, a gorilla kicks a native in the stomach. In fact, the gorilla's anatomy makes such an act impossible.

fully behind bars. He has since been proved quite wrong.

He also declared that an attacking gorilla could be relied on to meet its enemy erect, to stand and fight and to kill with a powerful, slashing blow across the abdomen. Yet all he had to go on was the testimony of a poor native who stumbled across a well-hidden gorilla, wounded it in the side with a single shot and, finding himself too close to escape, paid the price for his carelessness. He suffered a severe abdominal wound and his gun was bent and flattened hideously. In the light of present-day knowledge, it is tempting to think that either the native shot himself in the panic (du Chaillu heard two shots, the first of which wounded the gorilla) or, having missed with his second shot, attempted to protect himself by hurling his gun at his wounded and enraged adversary.

Du Chaillu left a graphic account of this incident but the anatomists overseas were not impressed. They declared – and have since been proved correct – that the bone and muscle arrangement of the gorilla's hindlegs would not permit anything other than short bipedal excursions and were certainly not arranged in such a manner as to support the sustained attack of which du Chaillu spoke.

However, the greatest blow dealt du Chaillu was administered by Winwood Reade, distinguished American traveller. On the evidence of his own investigations, he declared that all horrific gorilla stories from Hanno onward were fabrication. He travelled extensively through Equatorial West Africa, retracing du Chaillu's footsteps and interviewing natives as he went. He learned that gorillas attack on all fours, and that they make physical contact with their enemies only if cornered, angered or severely wounded. And nowhere could he find any native who had known a human killed by a gorilla (although nearly all had heard stories to this effect). He became convinced that an attacking gorilla might inflict only a single bite before retreating, rarely anything more than that.

It was a turning point in the reputation of the gorilla. Winwood Reade may never have seen such an animal in the wild, but his investigations drew him to the conclusion that gorillas were more timid and secretive than had been supposed and that they were more frightened of man than he of them.

When he was in the heart of what we might call 'du Chaillu country,' he disguised himself as a tradesman and immediately was asked if he, as the white man before him, wished to purchase gorillas. In this he refused but added that he would pay a handsome price to the man who could lead him close enough to a wild gorilla to get a shot at it. The natives, to Winwood Reade's delight no doubt, were astonished, enquiring as to why he should want to do the one thing other white men had not wished to do. With his interpreters he interrogated six natives who had accompanied du Chaillu on his forest excursions. They spoke affectionately and highly of du Chaillu, but all vowed that he had never shot a gorilla. If so, the question remains: how and where did du Chaillu procure the specimens he sent to England? Possibly the natives were themselves responsible for the deaths, whereupon they led du Chaillu to his dreams. Or perhaps he *was* telling the truth but the natives would not admit to playing any part in the killing.

The second half of the 19th century was the heyday of African exploration. John Speke, James Grant, Verney Cameron, David Livingstone and Sir Henry Stanley covered the central mountainous region, all eager to discover the source of the Nile. Only a few of them suspected the existence of a discreet population of gorillas tucked safely away in the densely forested

American showman Robert Noell performs with his massive pet, Tommy, in a routine that cashes in on the gorilla's perennial King Kong image.

Virunga Volcanoes.

Most of these men must have seen mountain gorilla country both near and afar. Speke named the distant volcanic cones Mfumbiro Mountain – only later was the named changed to Virunga – and even heard of monstrous men that could not talk. The word gorilla is to be found in several reports from this area, but there is little specific information.

At the turn of the century, a certain E Grogan, on an epic march the length of Africa, stopped off at the Virunga Volcanoes to dally in big-game hunting. He found (and left lying there) the skeleton of an ape so large that it could never have been a chimpanzee, an ape with which he was very familiar. In 1902 the German Captain Oscar von Beringe began an assault on Mt Sabinio, one of the Virunga peaks. At an altitude of just over 9000 feet, he and his companion, a Dr England, shot and killed two large, black apes. Of one thing they were certain: if these were chimpanzees, they were of a size hitherto unrecorded.

During the first 20 years of the 20th century, there followed a great deal of hunting both for sport and for museum collections. Specimens were passed from institute to institute, from taxonomist to taxonomist. The literature was flooded with descriptions of several kinds of gorilla on the evidence of only limited material. The assumption made was that the gorillas lived in small, isolated groups with no flow of genetic material between them. Consequently the differences between them – today recognized as no more than variation between genetically-linked individuals in such respects as sex, age and size – were attributed to separate evolutionary forces. It was an attempt to identify these forces that led to the erroneous recognition of no less than nine different forms of mountain gorilla by the end of the second decade of the century.

A mountain gorilla strips bamboo, a favorite food, which grows at 7500 to 10,000 feet altitude. Gorillas customarily break off the young bamboo shoots and chew them to get at the soft center.

Snowflake, the Albino Gorilla

Snowflake, the only albino gorilla ever recorded, is the object of much curiosity and excitement in Barcelona Zoo, where he has been living since his mother was shot in the wilds of lowland West Africa for raiding a banana plantation.

Snowflake – his Spanish name is Copita de Nieve – was removed to Europe, where human foster parents nurtured him through his early vulnerable years. Once he had matured, he mated successfully with two normally colored females, producing a daughter and a son.

Although neither of these is an albino, attendant scientists are hoping that one day Snowflake will sire a youngster of a similar color to perpetuate the genetic mutation that makes the father such a prized possession.

Since then the number of species and varieties has fluctuated according to the amount of material examined and the opinion of the person conducting the examination. Today three subspecies are recognized, the Western Lowland Gorilla *(Gorilla gorilla gorilla)* from the Congo River Basin, the Mountain Gorilla *(Gorilla gorilla beringei)* from the Virunga Volcanoes and nearby Mt Kahuzi, and the Eastern Lowland Gorilla *(Gorilla gorilla graueri)* from the surrounding inland area. Until 1970 the latter two were grouped under the single name of the mountain gorilla, but extensive examination of a number of skulls revealed differences worthy of taxonomic recognition. In general all three are similar, with the Eastern Lowland population forming something of an intermediate between the other two more clearly separated groups.

We have come a long way since those first myths of animal brutishness. It may seem surprising that the first-discovered and more widely distributed lowland gorilla is today the least well known of the geographically isolated populations. Perhaps the mountain gorilla happens to live in such an exotic outpost that explorers were drawn magnetically into its domain.

Once there, it is true that many of them overindulged in the unnecessary killing of wildlife, but men of compassion and a true sense of conservation were also drawn into the arena. One such person was Carl Akeley who

Face to Face with Gorillas

'The gorilla group was straggled out below us. Mushamuka, the great silverback leader, lay in a clearing, glancing at us suspiciously but not sufficiently worried to raise his four-hundred-and-fifty-pound frame from its resting position. A few feet away a female carrying a tiny black baby had uprooted a small bush and was carefully divesting the twigs of their bark before transporting them leisurely into her gaping mouth. The rest of the group was scattered over about seventy yards. Movements among the branches and herbs showed where animals were still feeding or juveniles were playing. We could hear occasional rustlings, twigs snapping, coarse scratching noises and even Mushamuka passing wind. Down the hill to our left an animal repeatedly beat its chest but the displayer was hidden from our view by the trees. The group was resting after its substantial morning forage. The animals would remain relatively sedentary for the next few hours.

Two baby gorillas climbed up a willowy tree just twenty-five yards away. One infant sat in a fork swaying the tree gently to and fro, clapping his little hands and scrutinizing me with bright button eyes. The other thumped her tummy with one fist then, to improve her infantile display, picked a bunch of leaves and, with them clenched between her teeth, continued her bizarre performance.

The tree was obviously not going to take the weight of both animals and the second sensibly descended. Perched in a crotch about twenty-five feet above the ground the remaining infant swung the tree about. His playful girl-friend climbed up into a bush immediately behind him. She was slightly larger and fluffier with a reddish tinge to her dark coat. She stood upright in the top of the bush, beat her chest then flopped down, a shaggy black bundle. Her playmate still swayed backwards and forwards on his slender mast. He hung by one arm and one leg, smacking his hairy chest with the free arm, then somersaulted over with casual ease. Another, even smaller, baby climbed up and joined the reddish female in a game of thumping and tumbling. The larger infant stood upright and her new companion came up behind placing an arm on her back to form a tandem. The female lunged forward and grabbed the black male as he swung past on his sapling. She dragged him off into the top of the bush and all three youngsters somersaulted and grappled delightfully in front of us.

Mushamuka rose purposefully to his feet. He turned his calm, blemishless face towards us, then headed off down the slope into the trees. The incredible breadth of his great silver-haired back rocked solidly with every step. He was a truly magnificent gorilla. The female with the baby followed after him, the infant clinging on to her back like a tiny jockey on a gargantuan mount. Just short of the trees the mother stopped and lay watching us with a disinterested gaze. The infant was fascinated, however, staring at us with thumb in mouth. A young juvenile gorilla climbed slowly into view up a slender muvula tree and, staring playfully at us, thumped his little chest in mock threat. Climbing higher, he began twisting and cupping the branches to make a play-nest.

It was wonderful to see the gorillas at such close quarters and so unbothered by our presence. Some animals are an anticlimax when you finally see them in the wild: they look no more exciting than in a park or zoo; but not the gorilla. The gorilla is a truly impressive animal with a very real and powerful presence. Normally gorillas are shy of man and would not permit such close inspection. The fearlessness of Mushamuka's group was a tribute to several years of courageous perseverance by a one-time Belgian tea-planter. Adrien Deschryver had visited the two gorilla groups residing in the forest beyond his plantation for six years and by his quiet, calm and unaggressive manner he gradually gained their acceptance. Getting the gorillas to transfer their trust of him to other humans was another problem, but eventually the Kahuzi-Biega became a National Park with African staff trained to take visitors or scientists out to see the gorillas. Tolerated as we were, however, gorillas are not animals with which to fool around. I could see that my guides were frightened, extremely cautious in their visited the volcanoes in 1921 to collect gorillas for the American Museum of Natural History. Although he came away with five specimens, he sensed the dangers of indiscriminate shooting in such a restricted and impressive site and turned his attentions towards the Belgian Government which controlled the area. In 1925 politics paid homage to science and the Albert National Park was established. Initially it contained within its boundaries only three of the majestic Virunga peaks, but four years later it was expanded to embrace the whole of the volcanic chain.

In 1926 on a return journey to study the gorillas, Akeley died and was buried high up between the two peaks of Mt Mikeno and Mt Karisimbi. I stood, just once, on the high grounds of the Ruwenzori Mountains in Uganda more than 100 miles to the north and gazed southward – across a rough sea of green, its peaked horizon obscured by rolling mists – and realized at first-hand the extent of his achievement.

In the 1930's the Virunga Volcanoes, in which hunting was prohibited, provided about the only safe place for mountain gorillas to live. Elsewhere their numbers were being reduced substantially, not only through shooting but also through live trapping for export to foreign zoos. The most despicable method of obtaining live specimens of a manageable age and size was – and may still be today – that of shooting an adult female and then removing her

every movement. They did not enjoy our proximity to the great apes and came into the forest only because they were paid well to do so. Their fear was not without foundation, for when forest pygmies still hunted gorillas with bows and spears many of them died at the ferocious hands and jaws of the enraged apes. Adrien Deschryver once found the remains of a pygmy who had been torn apart by a gorilla and strewn about the bushes. There is a widespread belief that if you stand and face a charging gorilla you will be quite safe. Curiously enough for Europeans this maxim seems to hold. European scientists like Deschryver have found that the gorilla will not attack a person who stands his ground, but the image of the gentle giant is as misleading as the older idea of the savage killer. To make a threatening move in the proximity of wild gorillas is a sure way to provoke a charge, and to then show fear by running away could invite a real attack.

It began to rain, just a drizzle, but the gorillas moved into the shelter of thick bushes and there was little sign of activity. We crouched among the broken herbage and I drew my cape over my camera. The rain continued for twenty minutes or so before the clouds parted and thin streamers of sunshine pierced the gloom on the other side of the valley.

After twenty more minutes the clouds drifted away over the ridge and the whole valley was bathed in sunshine. Wisps of steam rose from all sides. A gorgeously-coloured bee-eater flitted among the branches above us, launching down to snatch insects with a loud click of his beak then returning to his perch to eat his prizes at leisure. Clouds of tiny flies rematerialized and a gaudy Charaxes butterfly hovered with stiff wings before settling on a fresh gorilla dung to probe this apparent delicacy with its long proboscis.

There was no sign of the gorillas and I was afraid they had moved off down the hill unseen and unheard in the rain. We decided to follow their trail and set off cautiously down the slope. My tiny guide, Patris, led the way and I helped him flatten the herbs in front of us while the other guide hung back to watch and listen for any tell-tale movements ahead. The theory behind this method of progression is that although gorillas will charge right up to humans if they have vegetation around them, they seem shy of charging out into the open. We reached the bushes where the gorillas had feasted earlier, and followed their narrow track through a gap between two dense thickets. Patris and I had just entered a tiny clearing where the herbage had been crushed by lazing gorillas when there was a grunt ahead and our lookout behind us yelled *"Ngila! Ngila!"* (gorilla).

Terrific barks, followed by a crashing charge, and Mushamuka is screaming at us from only a few yards distant, hidden by the bushes. We can see nothing and Patris tries to run back along our path, but I grab him and hold him as it would be dangerous to pass through the thicket so close to the angry silverback. Patris struggles but I refuse to let go as the roaring gorilla charges towards us. He is huge, a mountain of powerful muscle. We stand facing him as he dashes to within four yards and then stops in the clearing, screaming, tearing up herbs with his enormous arms and hurling them into the air. My heart pounds and I am dizzy with fear, but remain absolutely still, gazing into his eyes. I can see the terrible scarlet lining to his yawning mouth and his great white teeth; his powerful, musty odour fills the small clearing. Suddenly he relaxes, closes his mouth and looks completely calm. He turns and ambles back into the thicket, and we creep silently back along the path to the other guide, shaken by our experience. We had been wrong in thinking the gorillas had left. Obviously they had been resting in the thicket and were most disturbed by our rude interruption.

Patris was furious at my behaviour. In front of his sympathetic companion he mimed a graphic repeat performance of Mushamuka's charge and how I had prevented him from running away. He was sure I had wanted to make an offering of him to the angry gorilla and it was several days before I could persuade him that I had acted in the interest of our common safety.'

John MacKinnon, *The Ape Within Us*

youngster as it clung pitifully to her dead or dying body. Not all came to shoot or trap, however, for the camera and sheer curiosity were becoming respectable enough motives for travel.

In fact there seemed to be such an increase of visitors to the area before World War II that it is intriguing to speculate how important a part the outbreak of war played in affording the gorillas something of a respite. The literature, so abundant during the 1920s and the 1930s, ceased abruptly, and when it appeared again during the 1950s, the emphasis had shifted dramatically from killing to conservation. The seven years of human strife must surely have played its part in curbing the desire to slaughter such human-like animals.

Now the tide of fortune began to flow strongly in the gorilla's favor. In 1955 Walter Baumgartel took over the Traveller's Rest Hotel on the Ugandan side of the Virunga Volcanoes. He realized immediately the tourist potential of the location but, more than that, he appreciated the gorilla's immense value to science. His campaigning brought the gorilla and its plight to the attention of such people as anthropologists Louis Leakey and Raymond Dart. Before long, money was made available to support some of the fieldwork that would provide the answers to so many unsolved riddles. From 1956 onward studies of the mountain gorilla – the rarest of the three subspecies – gathered in momentum and many reports appeared in both the scientific and the popular literature before 1960.

There had not, though, been an intensive field study of gorillas. Such work would mean hardship, deprivation and a life of innumerable unknowns in a heavy and humid atmosphere. It would need patience, determination and the mind of the romantic coupled with that of the scientist. In the American George Schaller, the challenge found the very man.

In 1959 Schaller began what was to become one of the epic studies of a living animal. Although several notable people have enlarged his findings in the years that have elapsed since then, it is his work that provides the foundation for almost everything that is known about gorillas today. For the first time, a human being approached gorillas on their terms.

Disdaining firearms and suppressing fears, Schaller would move slowly and cautiously through the forests until he located a feeding party. He would present himself to them from a distance so that they could keep an eye on him and familiarize themselves with his presence. Gradually they would move closer and as they did, Schaller was able to note the way they behaved toward each other, their mannerisms and their moods. These were the tools of his chosen trade. By conforming to their codes of discipline, by playing second string to their behavioral bow and by adopting the role of a peace-loving creature himself, Schaller was able to elicit from those great animals behavior that only a few years beforehand would have been heralded as ridiculously impossible.

By combining his two years of observations with reports from reliable natives and the literature available to him, Schaller was able to build up a picture that was to do much to set the record books straight. He and his wife Kay emerged from the jungle with an astonishingly detailed appraisal of the apes – enough to dispel the King Kong image for ever.

We now know enough about gorillas to appreciate their gentle manner, their strictly vegetarian diet, their surprisingly low sex drive and, above all, their total unwillingness to mount an attack unless severely provoked. The fearful displays of chest-slapping, of barking screams and of frenzied vegetation-tearing are all economically evolved to keep intruders, par-

An old silver-back peers out of the foliage at the intruding photographer.

ticularly human ones, at bay. The more persistent the intruder, the more likely is the display to become an attack, but even then only rarely will physical contact be made.

Schaller was able to approach his subjects closely because he made it his responsibility to pay homage to their life style. He crossed the threshold of centuries of ignorance and fear, entering a privileged world in which he, sole representative of the human species, was accepted. Perhaps the biggest barrier that he crossed was that imposed by the gorillas' threatening behavior. He noticed that one particular gorilla full of curiosity approached him, staring, and began to shake its head when at a distance of some 20 yards or so. Schaller later reciprocated the gesture and the gorilla immediately looked away, thereby avoiding visual contact. Was this a real attempt at friendly communication, threatening stare being counteracted by a head-shaking gesture of submission? When next Schaller stumbled across a lone male at close quarters, he immediately shook his head. It *was* a universal gesture, for the gorilla, apparently satisfied that the 'ape' before him meant no harm, turned and ambled off into the forest.

The lives that gorillas lead in some ways do appear to be at odds with the potential suggested by their physical strength and bulk. Man apart, they have no enemies save the occasional leopard. They probably evolved to their size when large predators were more numerous and unlike the smaller chimpanzee, gorillas were more restricted to the ground. The body that gives such protection has to be fuelled constantly. The jungle provides a constant supply of fuel, a far greater and more reliable supply than animal prey. But vegetation is far poorer in terms of calories and protein than meat, and gorillas must feed for many hours of the day if they are to satisfy their bodily requirements. They lead an idyllic existence – waking, feeding, resting, feeding and sleeping, moving slowly through the thick undergrowth.

Within this rather easy-going framework, however, each group is strictly dependent upon the whims of one adult male: the grand old silver-backed individual around whom all group activities revolves. It is he who decides where the group – numbering from up to 30 strong – shall go to forage during the day, he who decides how long they shall stay there, when the day's activity is to cease and when the ritualized nest building is to begin. It is the responsibility of every individual to respond to his movements and emotions and once his intentions have been signalled, they spread rapidly through the group, which acts as if united by a single word of command. It is he who leads the way to a new feeding ground, while the others – typically an array of females with infants, less dependent juveniles and the maturing black-backed males – string along behind.

Sometimes when progress is more leisurely and perhaps not seeking to break unfamiliar ground, the silver-backed male is to be found ambling along in the middle of the group. But the hierarchy remains, albeit temporarily relaxed. The ranking is so well organized that gorillas rarely squabble and fight among themselves. The silver-back reigns supreme and there may even be other silver-backs under his command. All of these individuals preside over females and the younger black-backed males who, in turn, are dominant over all juveniles and infants which are not being closely attended by their mothers. If they are, then mother and child are elevated in a kind of status juggling that maintains group cohesion while affording maximum safety to the new generation.

Thus cushioned against the dangers of the outside world by a strategy promoted by natural selection, the infant gorilla is embarked upon a long

Although shot in a zoo, these two pictures show the closeness that is typical of mother and baby in the wild. Despite the huge disparity of size, the mother plays with the baby and nurses it with a tenderness displayed by all apes, ourselves included.

and vital trail of maturing and learning. Schaller has pieced together the typical life-pattern of a newborn gorilla over the first six months of its life.

The picture is one of a helpless creature showly gathering its physical and mental faculties until capable of its first exploratory steps into the world of its peers. During the first four or five weeks it is so powerless that it cannot even support itself against its mother's body unaided for more than about ten seconds at a time. By two months, however, it has progressed quickly and is sufficiently well developed to ride on her back, its tiny hands clutching firmly while an inquisitive head is raised periodically to gaze at the mysterious enormity of its surroundings. Month three finds the infant increasing its physical range, crawling beyond its mother's body: vines and grasses may be grasped, pulled and even bitten into. The new playground may also encompass the bodies of other gorillas, all of whom are explored and tested under the protective gaze of the mother who sits attentively nearby, instructing by not interfering with this harmless probing. Schaller noted the first real signs of adulthood when, between four and five months, a youngster, walking on all fours, raised its body erect against the stem of a lobelia plant and, in a shaky two-legged stance, began to beat its chest. By the arrival of the sixth month of its life, the young gorilla is strong enough to run around on all fours and even to haul itself up into small bushes and trees.

At this stage the zestful infant is drawn by increasing activity and curiosity into more contact with the variously-aged gorillas of its group. While it is tolerated by those to whom its antics present no status or sexual threat, it soon learns to respect the moods of individuals. Some are less tolerant than others, demonstrating their annoyance at being pestered by a single glance or a persuasive cuff of the hand. But others may allow themselves to be carefully examined, paying little heed to the detailed inspection of the contours of their bodies.

In general the old silver-backs take little notice of the youngsters who often seek their company specifically, unless the young ones have strayed too far from their mothers and need quick protection. Schaller once noted an aging male which, when his group accidently came across another group of gorillas at close quarters, scooped up a youngster and ran with it tucked under his arm for some 10 yards before setting it down again.

Unlike the silver-backs whose status demands that they go along with the hero worshipping of the very young, the middle-ranking black-backed males seem to be rather by-passed in the juveniles' hunt for idols. Occasionally they meet and temporary bonds are formed, but normally the two age groups pass their maturing days with little to do with each other. When the youngster is not learning about hierarchies and their importance to group welfare, it is experimenting with its own food gathering.

As early as two or three months the young gorilla is beginning to eat solid foodstuff in addition to the milk obtained from its mother. Whether it is dependent upon this intake or whether it just forms part of its curiosity to pull and pick at its surroundings is not certain. Much of this behavior probably derives from direct observation of its mother during her extensive daily feeding bouts. Gradually the dependence upon milk is relaxed and the diet of various grasses, herbs, vines, sedges and shrubs assumes increasing importance. The bark and pith of some trees may also be eaten.

The infant's mother apparently pays little attention to the foodstuff her offspring is experimenting with, although Schaller did notice occasions when co-operation played a significant helping role. One youngster

A male resting on the ground.

A rare sight: mountain gorillas in a tree.

London Zoo's Ill-Fated Guy

For thirty years, Guy the Gorilla was London Zoo's biggest attraction. Tragically, his very popularity proved to be his final undoing.

He was picked up in the French Cameroons in 1946, when he was just six months old. After spending a year in the Paris Zoo, he came to London on the 5th November 1947 – Guy Fawkes Day, hence his name. His immediate popularity brought him a positive cornucopia of gifts from the public – strawberries, grapes, nuts, melons, pineapples and ice creams, all of which he ate with delight. By the time the Zoo put a stop to public feeding in January 1968, he was overweight and his teeth were beginning to decay.

Thereafter his health improved and his list of admirers grew. Many were such regular visitors that Guy knew them personally and could spot them at a distance in a milling summer crowd. He has no progeny; by the time a female gorilla was available, he was, in the words of a zoo official, 'a staid old bachelor.'

In 1978, when he was 32 and the oldest gorilla in Europe, he seemed in good shape. He tipped the scales at 380 pounds with a chest measurement of 73 inches – by no means a giant among gorillas. But his calorie-rich diet caught up with him. Some of his bad teeth began to pain him and it was decided to remove them under anaesthetic. The strain proved too much. During the operation he had a heart attack and never regained consciousness.

actually removed a piece of the vine *Galium* from its mother's mouth and ate it itself. Another was chewing some rather unsavory *Hagenia* leaves when a female approached and removed them from its hands. One other youngster was struggling to uproot the firmly embedded stem of a lobelia plant with such difficulty that a female watching nearby finally moved over, snapped off the stem and left it on the ground. The youngster then picked it up and began to chew it.

The overall impression is of infant gorillas exploring within a close-knit group whose members show varying degrees of response to them. All show concern for their welfare. While the young are permitted to investigate all aspects of social life so that they are familiar with all its facets, the mother of each of them remains as the focal point of their development. She will rescue them in moments of distress and always be available should her offspring get into trouble or feel a need to return to the warmth and security of her comforting frame.

By the time the youngster is well into its second year it is almost completely weaned and is able to move quite freely within the travelling group, perhaps hitching a ride on the back of an older female if progress is too quick for it. Each evening it settles down with its mother in the temporary nest she has constructed either on the ground or in the lower reaches of a bush or tree. Such behavior continues for up to three years while the youngster absorbs the way of life around it.

By its fourth year its increased weight and independence make such reliance upon another individual less possible and it is left to its own devices. By its sixth year it is fully independent.

If it is a female, this independence marks the onset of sexual maturity and she may well conceive soon after, thus to repeat every few years the pains-taking process of mothercare that brought her so successfully through her formative years.

But if our growing youngster is a male, it takes its low-ranking place as a maturing black-back. At about ten years, and now approaching sexual maturity, the masculine features of a pronounced sagittal head crest, strong muscular body and a detectable graying of the hair on its back, legs, neck and even sides, announce that it is at last developing into one of the world's most respected animals – a high-ranking silver-backed male of phenomenal strength. When it stands erect it may reach over 6 feet in height, supporting on its massive leg muscles a solid bulk of up to 500 pounds. It may live for 30 to 40 years.

Gorillas, of course, have been living like this for rather a long time. Long enough in fact to have become highly specialized and dependent upon what is now a fast-decreasing habitat in Africa. If the gorillas are to survive they will do so now only as a result of careful human management. People like George and Kay Schaller have pointed us in the right direction. A notable successor to the Schallers has been Dian Fossey who is still toiling almost singlehanded in a battle against politicians, agriculturalists and poachers. Indeed the gorillas seem much closer to her in spirit than her human opponents. Once she stretched out a friendly hand toward a young male gorilla whose trust she had gradually won. He hesitated, stretched out his own hand, hesitated again and then gently touched her fingers with his own: the gulf of centuries crossed in a brief moment of confidence. We have come a long way in our understanding of gorillas in the last century. Who, one is forced to ask, now seems the more monstrous? Them or us?

CONFRONTATION WITH A SILVER-BACK

Perhaps the most remarkable incident connected with the study of gorillas – and certainly the most remarkable recorded on film – occurred in in the Kahusi-Biega National Park in Zaire in 1974. The park's director, Adrien Deschryver had, over a number of years, built up a personal relationship with the dominant male of one of the family groups, a gorilla he called Kasimir. Kasimir had come to recognize Deschryver and knew him to be no threat.

Part of Deschryver's job, as director, is to care for any orphan gorillas. In 1974 he heard of a six-month female baby whose mother had been killed by poachers. Deschryver picked up the baby, reared it by hand and then, as part of an attempt to acclimatize it once again to forest life, decided to test its reactions in the presence of adult gorillas in the forest. The incident that followed was filmed by the Survival unit of Anglia TV and incorporated in a classic wildlife film, 'Gorilla.'

Accompanied by photographers Dieter Plage and Lee Lyon, Deschryver, carrying the baby, ventured out in Kasimir's territory. No sooner had the three found the group than they realized that they were surrounded by some 20 gorillas and that retreat was impossible. Apparently incensed by the cries of the baby gorilla, Kasimir false-charged a number of times and then burst out of the bushes to confront Deschryver face to face. Deschryver, convinced he would otherwise have been killed, dropped the baby. Kasimir swept it up, retreated into the bushes and presented it to his family.

Tragically, his family contained no lactating females at that time and this fact, in combination with a bout of cold weather, led to Julie's death shortly afterward.

Deschryver watches Kasimiar slap his chest – a sign not of aggression, as often supposed, but of uncertainty.

Feeding, Kasimir displays the massively broad silver back that marks him as a mature male.

With a sidelong glance at the familiar figure of Deschryver, Kasimir ambles across the line of the camera and back into the forest.

Kasimir's immense bulk looms through the mountain forest in which he and his family group live and feed.

Deschryver's wife bottle-feeds the orphaned gorilla baby, Julie.

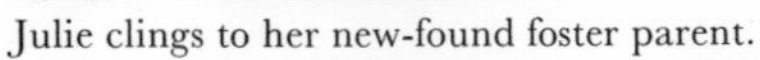

Julie clings to her new-found foster parent.

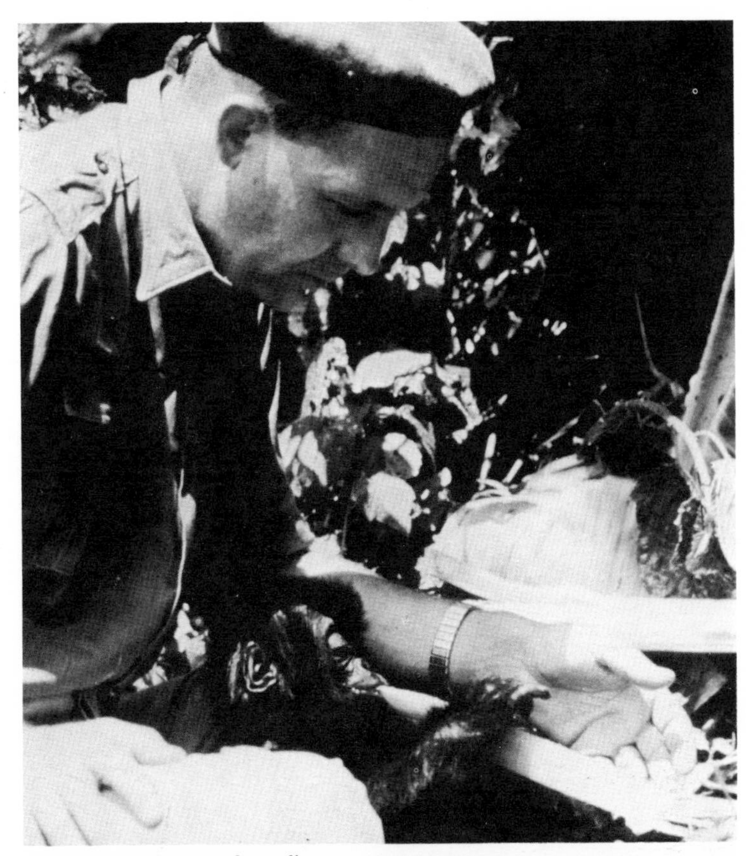

Julie is weaned onto a forest diet.

Kasimir bursts from hiding to confront Deschryver and seize Julie.

4/CHIMPANZEES: THE SOCIAL EXTROVERTS

The chimpanzee is man's closest relative (though exactly how close is still hotly debated). Chimpanzees live in and around the tropical rainforests of Africa from Sierra Leone to Tanzania (below), where their basic diet of fruit, nuts and young leaves – with occasional feasts of meat – is available all the year round. Although often displaying great individuality of gesture and character (left) they are, above all, social animals, living in changing groups of 30 to 80 individuals.

One of the more confusing zoological issues at the beginning of the 19th century concerned the chimpanzee and the gorilla. Chimps had for a century been quite well known from the coastal regions north of the Congo River (now called the Zaire River), and from Angola (the name chimpanzee is itself Angolan), when from further inland there came news of an ape of ferocious strength and brutal habits. The existence of the gorilla had still to be proved and scientists, assuming they were one and the same, expended considerable time in attempts to reconcile an ape of the chimpanzee's known proportions and temperament with the abominable ways attributed to the inland apes.

In 1817 the great French anatomist, Baron Georges Cuvier, writing of the chimpanzee in his *Animal Kingdom* reveals the confusion between the chimp and gorilla (the brackets were added later):

'Covered with black or brown hair, scanty in front; (a white marking on the rump). If the reports of travellers can be relied upon, this animal must equal or be superior in size to Man. (The skeleton of an adult female in London is considerably smaller.) It inhabits Guinea and Congo, lives in troops, constructs huts of branches, arms itself with clubs and stones, and thus repulses Man and Elephants; pursues and abducts, it is said, negro women (very highly improbable), &c. Naturalists have generally confounded it with the Ourang-outange. In domestication it is very docile, and readily learns to walk, sit and eat like a man. (It is much more a ground animal than the Ourangs, and runs on its lower extremities without difficulty, holding up the arms. It is of a lively and active disposition. . . .)'

Despite his towering reputation as an anatomist, Baron Georges Cuvier perpetuated the 18th-century confusion between chimpanzees and gorillas. The existence of the two separate species was confirmed only in the mid-19th century.

The first gorilla specimen, a skull, was discovered in 1847 and the chimpanzee was largely exonerated of the legendary crimes that had been heaped upon its shoulders. The international excitement generated as a result of the confirmation of the gorilla's existence forced the chimpanzee somewhat into the scientific background, but information about it grew steadily. As explorers like Paul du Chaillu and David Livingstone penetrated the interior of Africa, they discovered that the chimpanzee enjoyed a wide and varied distribution in tropical Africa, showing adaptations to dense forest on the one hand and to more open savannah woodlands on the other. The explorers came across populations of chimpanzees showing such differences in morphology and behavior that the literature was soon flooded with descriptions of new kinds of ape.

It is worth examining the twists and turns of these discoveries and the related taxonomic confusions, for they remind us how difficult it once was to identify little known animals correctly. At one time no less than 20 species of chimpanzee were recorded. There is still some controversy over how to classify the several different forms.

Du Chaillu's dream was to unravel the mystery surrounding the gorilla and it is in respect of this ape that he is largely remembered. But his contributions toward a better understanding of the chimpanzee deserve to be brought more fully into the open.

In 1855 du Chaillu pushed his way eastward through Gabon. Sometime later he discovered a rather strange looking shelter built on the branches of a forest tree. His native porters assured him that it was not, as he suspected, the overnight home of a hunter but that it had been constructed by a *nschiego mbouvé*, a kind of ape – a curious creature with a bald head – not unlike the fabled gorilla they were now searching for.

The party pressed on and before long encountered more of these nests

This alleged portrait of a chimpanzee captured in Angola in 1738 represents the animal as little more than a monkey-faced human. The artist was unable, apparently, to accept that chimpanzees do not walk upright by preference and that their stooping gait is not a sign of infirmity that demands a stick.

CHIMPANEZE, agé de 21 mois haut de 2. pieds 4. pouces
apporté d'Angola en 1738

Chimpanzees in Kivu Province, Zaire, relax on the exposed branch of a dead tree. Reaching a maximum weight of only about 150 pounds, chimpanzees are well adapted to a life both on the ground and aloft.

ranging in height from 15 to 50 feet above ground, in response, du Chaillu reasoned, to the relative abundance of the carnivorous leopard. The natives said that the nests were elaborately built by the male *nschiego* while the female collected the vine and twig materials with which he wove the temporary home. When satisfied with his work, the male allowed the female to enter before constructing his own shelter nearby. These shelters were never occupied for more than about ten days at a time, for the *nschiego* lived on wild berries and when its supply was exhausted it moved on to a new locality.

Du Chaillu never saw many nests together and he concluded that the ape lived not in troops as had been reported for coastal chimpanzees, but singly or in pairs. His natives said that a solitary nest inhabited by a single *nschiego mbouvé* with silvery hair denoted its advanced years concommitant with a desire for solitude and peace after a long and troublesome life.

Making nests is an integral part of chimpanzee behavior. They are mostly made as nighttime beds in trees (though they may on occasion be made on the ground and for a rest during the day). The first stage involves selecting a secure place in a tree (left), then bending fronds and branches down into position and inter-weaving them (below left). The result (below) is a secure bed, safe from predators. The nest is always kept clean and chimps are careful to defecate and urinate over the edge.

Sidetracked by curiosity, du Chaillu decided that he should try and collect one of these strange apes, for they were obviously new to science and worthy of a small amount of his time as he searched for the gorilla. He planned to find a fresh-looking nest and then to wait and see if an ape returned to it as evening descended on the forest. His success in this ploy was instantaneous for, '. . . we at last, just at dusk, heard the peculiar "Hew, hew, hew," which is the call of the male to his mate. We waited till it was quite dark, and then I saw what I had so longed for all the weary afternoon to see. A Nschiego was sitting in his nest. His feet rested on the lower branch, his head reached quite into the little dome of the roof, and his arm was clasped firmly round the tree-trunk. This is their way of sleeping. After gazing till I was tired through the gloom at my sleeping victim, two of us fired, and the unfortunate beast fell at our feet without a struggle or even a groan. We built a fire at once, and made our camp in this place, that when daylight came I might first of all examine and skin my prize. The poor Ape was hung up to be out of the way of insects, and I fell asleep on my bed of leaves and grass, as pleased a man as the world could well hold. Next morning I had leisure to examine the Nschiego.

'I was at once struck with points of difference between it and the chimpanzee. It was smaller, and had a bald black head. This is its distinctive character. This specimen was three feet eleven inches high, or long. It was an adult. Its skin, where there is no hair, is black, and the thick breast and abdomen are covered with short and rather thin blackish hairs. On the lower part of the abdomen the hair is thinnest, but this is not perceived

The chimpanzee's foot bears a structural resemblance to the human hand, with a big toe that is offset from the other four digits and fully opposed, so that it can be used for climbing and grasping. But also like the human hand, it can, if necessary, be placed flat on the ground (as is clearly visible in the shot below of two chimpanzees in characteristically human poses). The chimpanzee does not, therefore, have to adopt the ungainly gait that characterizes the orang.

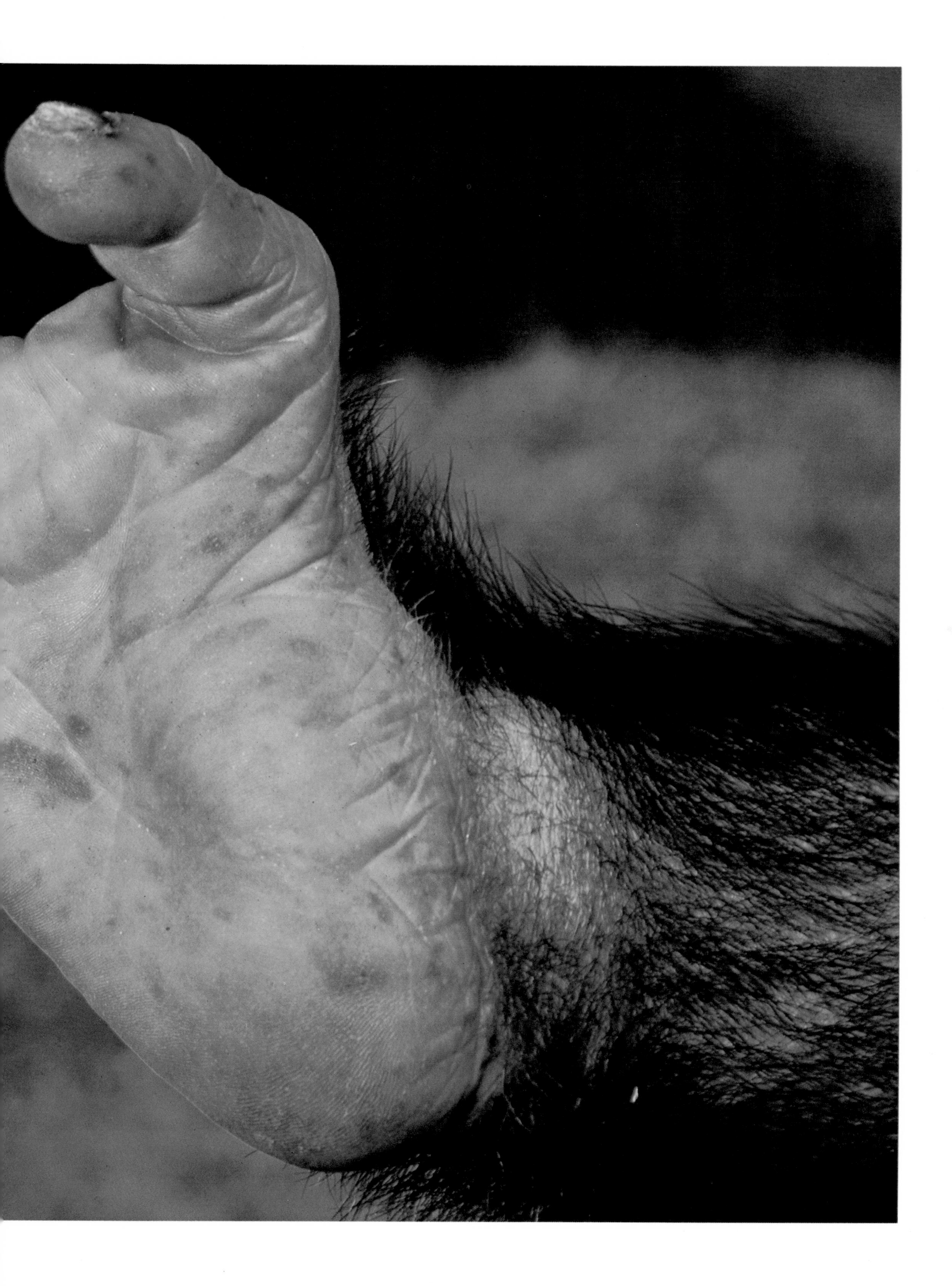

unless looked at carefully, as the skin is the colour of the hair. On the legs the hair is of a dirty gray, mixed with black. The shoulders and back have black hair between two and three inches long, mixed with a little gray. The arms down to the wrist have also long black hair, but shorter than in the gorilla. The hair is blacker, longer, glossier, and thinner in general than that on the gorilla, and the skin is not so tough. I noticed that the bare places, where the hair is worn off by contact with hard substances in sleeping, were different from the bare places which are so conspicuous on the common chimpanzee.'

Evidently du Chaillu collected his first *nschiego* specimen after he had been able to examine a gorilla – presumably a dead one – at close quarters. He makes other pertinent comments on the differences between the two apes. Of the *nschiego* he says, 'It is not as powerful an animal as the gorilla, its chest is not so large, but the arms and fingers are a little longer, and this is the case with the toes also. The nose is not so prominent, but the mouth is wider and the ears are larger. Its chin is rounder and has more small hairs.'

As du Chaillu became more involved with his new ape, so he devoted more time toward studying its behavior and habits. He compiled careful notes on the construction of the peculiar sleeping nests and even had his natives cut down a tree in which one had been built. On close inspection of the shelter, he found it exactly as he had described it from ground level, 'of long branches and leaves, laid one over the other very carefully and thickly, so as to render the structure capable of shedding off water. The branches were fastened to the tree in the middle of the structure by means of wild vines and creepers, which are so abundant in these forests. The projecting limb on which the Ape perched was about four feet long. There remains no doubt in my mind that these nests are made by the animal to protect it from the nightly rains. When the leaves begin to dry to that degree that the structure no longer throws off water, the owner builds a new shelter, and this happens generally once in ten or fifteen days. At this rate the Nschiego Mbouvé is an animal of no little industry.'

In 1855 Duvernay, a renowned French taxonomist, described this bald-headed ape as a new species, *Troglodytes tschego*, elaborating on all the features that set it apart from the gorilla. Nevertheless some scientists thought it merely a variety of the gorilla, differing only in size and anatomical details relating to a more arboreal way of life. Du Chaillu's own name for the species *Troglodytes calvus* appeared in July 1860 in the Proceedings of the Boston Society of Natural History.

At the same time du Chaillu described yet another erroneously named species of great ape, which he called the 'Koolo-Kamba' (*Troglodytes kooloo-kamba*), a name derived from its own forest call 'Koolo' and the native word 'Kambe' meaning 'to say.'

Du Chaillu had been alerted to its distinctive call and the fact that it inhabited the forests of Western Equatorial Africa alongside the common chimpanzee and immediately he set off in search of a specimen. He found just one in all his travels.

'We had hardly got clear of the Bashikoway ants and their bites when my ears were saluted by the singular cry of the ape I was after. 'Koola-koolo! koola-koolo!' it said several times. Gambo and I raised our eyes, and saw, high up on a tree-branch, a large Ape. We both fired at once, and the next moment the poor beast fell to the ground with a heavy crash. I rushed up, anxious to see if, indeed, I had a new animal, I saw in a moment that it was neither a Nschiego Mbouvé, nor a chimpanzee, nor a gorilla. Again

In this portrait a three-year-old chimpanzee looks questioningly at the camera with a gesture and expression that – as many zoologists now accept – can be justifiably compared to human reactions.

I had a happy day – marked for ever with red ink in my calendar . . . . The animal was a full-grown male, four feet three inches high, and was less powerfully built than the male Gorilla, but as powerful as either the Chimpanzee or Nschiego Mbouvé. When it was brought into Obindji, all the people at once exclaimed, "That is a Koola-Kamba." Then I asked them about the other Apes I already knew but for these they had other names, and did not at all confound the species. For these reasons I was assured that my prize was indeed a new animal; a variety, at least, of those before known.'

Other confusions followed. In 1866 the Koolo-Kamba was described from a different specimen as *Troglodytes aubryi* by Gratiolet and Alix, thus duplicating and confusing du Chaillu's *Troglodytes kooloo-kamba* of 1860.

While du Chaillu was plodding through the tropical forests of Western Africa, a new name had emerged from the east, nearly a thousand miles away, to seize the imaginations of the European world: David Livingstone, a legend in his own lifetime, devoted to missionary work and exploration, in particular to the search for the source of the Nile. He never found it, and died in central Africa in 1873. His African servants carried his body to the east coast and he was buried in Westminster Abbey, London.

Toward the end of his life, Livingstone encountered chimpanzees of a variety quite unknown to du Chaillu. They lived in more open woodland near lake Tanganyika and, like great apes everywhere, they played a significant part in the lives of the local natives. They called them *soko* (from the swahili for ape) but Livingstone (in his 1866-73 journals) also called them gorillas. Because of this, descriptions of them lay ignored – strangely, since gorillas were not known in the area – until one of his natives came to England and identified a stuffed gorilla as an animal completely different from the *soko* he knew so well. Livingstone's last writings therefore contain valuable information on woodland-dwelling chimpanzees.

In such areas, in contrast to the way of life in the rainforests, the natives' livelihood depends largely upon seasonal rainfall. Livingstone recalls a time when the rains were overdue and the natives were eager to capture a *soko* alive because its presence 'was believed to be a good charm for rain.' He continues, curiously, 'so one was caught; and the captor had the ends of two fingers and toes bitten off. The Soko, or Gorilla, always tries to bite off these parts, and has been known to overpower a young man, and leave him without the ends of fingers and toes.'

He also had the good fortune to see a *soko*'s nest and although he dismisses it rather scornfully as a 'poor contrivance – no more architectural skill than shown in the nest of our cushet dove,' it is a description well worthy of comparison with the elaborate structures fashioned by du Chaillu's *nschiego mbouvé* in thick tropical forest.

For all their future interest to science, however, Livingstone was obviously not attracted to his *soko*s. He found them ugly and not worthy of physical comparison with the lithe and supple figures of his natives or the antelopes of the plains. In one passage from his last journal he subjects the poor creature to all the venom he can muster from his ailing pen, weaving his own feelings into the native stories which so typically portrayed it as an animal intent on kidnap, rape and pillage.

'Four Gorillas or Sokos were killed yesterday; an extensive grass-burning forced them out of their usual haunt, and coming on the plain they were speared. They often go erect, but place the hand on the head, as if to steady the body. When seen thus the Soko is an ungainly beast. The most senti-

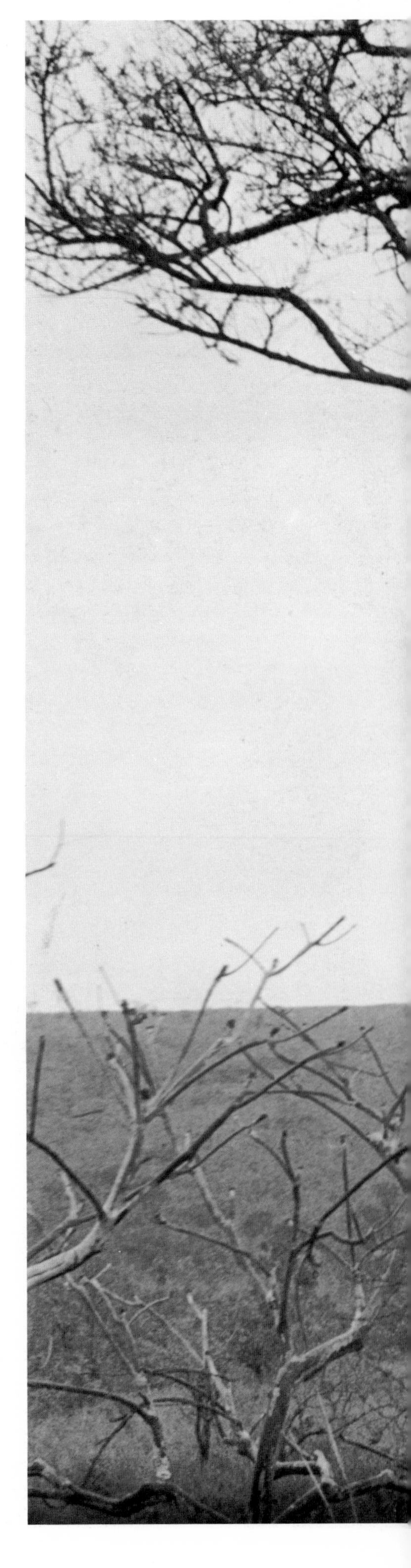

From the vantage point of bare branches, chimps crane their necks in the direction of an unfamiliar noise.

mental young lady would not call him a dear, but a bandy-legged, pot-bellied, low-looking villain, without a particle of the gentleman in him. Other animals . . . are graceful . . . but the Soko, if large, would do well to stand for a picture of the devil. He takes away my appetite by his disgusting bestiality of appearance. His light yellow face shows off his ugly whiskers and faint apology for a beard; the forehead, villainously low, with high ears, is well in the background of the great dog-mouth; the teeth are slightly human, but the canines show beast by the large development. The hands, or rather the fingers, are like those of the natives. The flesh of the feet is yellow.'

Having exhausted himself of his anti-*soko* feelings, Livingstone progressed into a description of *soko*-native relations. Livingstone's natives were largely tolerant of the *soko*'s lively antics. They obviously appreciated its intelligence and even allowed themselves to be subjected to many playful games until, perhaps, things got out of hand and they became frightened and defensive, knowing that beneath the facade there lurked a dangerous beast.

Livingstone continued, 'The Soko is represented by some to be extremely knowing, successfully stalking men and women while at their work; kidnapping children, and running up trees with them, he seems to be

In an incident reminiscent of the actions of children learning social behavior, a female reaches down in a begging gesture toward a bush-buck being consumed by her brothers. The two jealous males jerk away their prize and snarl a warning at their sister, who retreats out of harm's way.

amused by the sight of the young native in his arms, but comes down when tempted by a bunch of bananas, and as he lifts that, drops the child . . . one man . . . was hunting, and missed in his attempt to stab a Soko; it seized the spear, and broke it, then grappled with the man, who called his companions, "Soko has caught me!" The Soko bit off the ends of his fingers and escaped unharmed.

'The Soko is so cunning, and has such sharp eyes, that no one can stalk him in front without being seen; hence, when shot, it is always in the back; when surrounded by men and nets, he is generally speared in the back too, otherwise he is not a very formidable beast. He is nothing, as compared in power of damaging his assailant, to a Leopard or Lion, but is more like a man unarmed, for it does not occur to him to use his canine teeth, which are long and formidable. Numbers of them came down in the forest, within a hundred yards of our camp, and would be unknown but for giving tongue like Foxhounds: this is their nearest approach to speech. A man, hoeing, was stalked by a Soko, and seized; he roared out but the Soko giggled and grinned, and left him as if he had done it in play.'

Livingstone observed further that the *soko* never ate flesh and that it

An adolescent female uses a leaf for a delicate manicure.

David Livingstone, explorer and missionary, unwittingly provided detailed records of the behavior of woodland chimpanzees.

lived on wild fruits, preferring bananas above everything else. It would never touch maize. Occasionally they would collect together and make a loud drumming noise – 'some say with hollow trees.'

Much of Livingstone's information must have come second-hand from natives he visited and it is unlikely that in his declining years, confined to a sick-bed and for so much of the time deep in prayer, he witnessed much of it himself. If he was ever able to match his own observations against those of the natives, he was evidently swayed in his judgment by their beliefs in what they knew of the *soko*. Time and time again they assured him that a wounded *soko* always 'seizes the wrist, lops off the fingers and spits them out, slaps the cheeks of his victim and bites without breaking the skin.' Always the chimp is seen as the aggressor and never the victim of aggression responding in desperate self-defense. Even today in certain parts of West Africa, it is firmly believed that a *soko* will break a man's legs for no good reason at all.

The most interesting aspect of Livingstone's *soko* was its geographical range. It was widely held to be an animal closely related to the common chimpanzee, then called *Troglodytes niger*, and yet it was not associated with the dense tropical forest so typical of the species in the west of Africa. The only explanation at the time was that this was proof that once the great forests spread further to the east than they do today. This may well be a plausible explanation, but what is more arresting is the thought that Livingstone was witness to an adaptation from forest to grassland among a population of apes – an adaptation which many millions of years ago culminated in the evolution of man. Some interesting experiments, with significant results, have been conducted on forest and woodland chimpanzees to show certain evolutionary differences between them. These experiments will be enlarged upon later.

As regards the classification of chimpanzees, all the different names, both scientific and vernacular, which were bandied around over a period of 100 years or more have finally been resolved by fieldworkers and taxonomists. Today just two species are recognized. So far we have dealt with just one of them, *Pan troglodytes*, which is given three distinct subspecies or geographical variations. They are:

- *Pan troglodytes troglodytes*, the Chimpanzee (the *nschiego mbouvé* and the Koolo-Kamba), from the tropical regions of Gabon, Cameroons and the Central African Empire – occupying approximately that area of land between the Zaire and the Niger Rivers;
- *Pan troglodytes schweinfurthii*, the Long-haired Chimpanzee (the *soko*), from forested central Africa eastwards to the great lakes and woodland areas around Lake Victoria;
- *Pan troglodytes verus*, the Western Chimpanzee from the western bank of the Niger River westward through the forested countries of Ghana, the Ivory Coast, Sierra Leone and Guinea.

The chimpanzees mentioned so far occur mainly to the north of the Zaire River, though the subspecies *Pan t. schweinfurthii* also extends beyond the eastern headwaters of this great river and spills over to the south around the lakes of the western Rift Valley. There is, however, a second species of chimpanzee wholly confined to the forests of the southern bank of the Zaire, where it is a mighty river in full flow. This is the Pygmy Chimpanzee, *Pan paniscus*, or the bonobo as it may be more commonly known.

As late as 1929 the bonobo was described as a new subspecies of the common chimpanzee, but in 1934 it was elevated to a separate species. There will always be speculation as to the validity of such status, but the

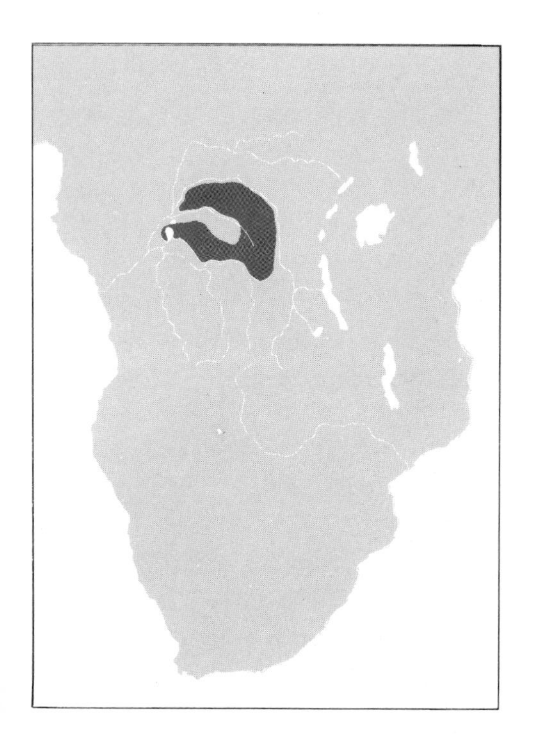

The bonobo, or pygmy chimpanzee, was recognized as a distinct species only in 1929. It is the only ape found south of the Zaire – Lualaba river system (see map, above). These smaller, lighter creatures are better adapted for travel in the treetops than chimpanzees proper. Their social life too is somewhat different. Chimpanzees form ever-changing groups and subgroups, but bonobos form smaller family units, more like those of gibbons.

bonobo remains today genetically isolated. It might, given the chance, interbreed with the other chimps, but does not do so in fact, and is thus at least an incipient species.

A combination of political unrest and natural remoteness saw the bonobo overlooked during the time that the pioneering field studies of the other apes were being conducted in the 1960s. More recently, during 1974 and 1975, Alison and Noel Badrian conducted a six-month survey of them in the deep forests enclosed within the Zaire, Lualaba and Kasai river systems. The initial impact of their preliminary study, combined with observations and speculations by John MacKinnon, suggests that the bonobo differs from other chimpanzees in ways other than just morphological. There is a strong and convincing indication that the bonobo is so restricted to a monotonous habitat by geographical features that it has become highly specialized, with little physical variation. By contrast its north bank relative has spread from the west coast to East Africa, adapting all the way to such an extent that morphological and behavioral differences are now plainly visible. There are, therefore, differences between the two populations well worthy of distinct taxonomic recognition.

In general bonobos are more arboreal than chimpanzees and spend many hours in search of widely-spaced feeding sites. They will often build their nests close to a fruit-laden tree where they can feed early in the morning and last thing before retiring at night. After their dawn feed they may travel more than a half a mile in search of the other foodstuffs which make up a balanced and well regulated diet. Apart from fruit and leaves, they will readily feed on animals including insects and small vertebrates. The Badrians found these remains in bonobo droppings especially prevalent between March and May when the usual supply of fruit was rather depleted. Supplementing a basic fruit diet with animal matter is an interesting adaptation, and one not readily found in the gorilla and orang-utan. It points to a new dimension among apes: one that was acquired by man.

The far-ranging behavior of bonobos and the small numbers of an average party make them difficult to observe in the forest. So, too, do their small size, quiet nature and habit of coming to the ground and making off through the dense vegetation. Their calls are screams, but are uttered only rarely – usually on the detection of a human intruder – and as a result the animals cannot be located as automatically as a band of noisy chimpanzees.

These points show some basic differences between bonobos and chimpanzees which have resulted from their ecological separation. The obvious difference, however, is that bonobos are about half the weight of chimpanzees, their bodies being smaller and more slender. They are also more agile in the treetops and their small group sizes (sometimes only four individuals) is more reminiscent of gibbons than of other apes.

The reason for these peculiarities lies in their isolation, with the Zaire River forming a dispersal barrier which has long prevented them from interbreeding with the north-bank chimpanzees.

The study of bonobos raises a fascinating question: which came first, the bonobo or the chimpanzee? Their close relationship implies that one is ancestral to the other.

This area of Africa is notoriously poor in fossil finds and there are no precise clues as to the phylogenetic relationships of the living forms. But their life styles, looks and distribution suggest a variety of answers.

One argument runs like this: gorillas and orangs are forest dwellers, bound to a typical habitat across their range; so are bonobos. Man is not

The relationship between the pygmy chimp mother and her baby is, presumably, as close as that between other primate mothers and their offspring, but the creatures are not yet well enough known for the details of parental behavior to be spelled out.

characterized by a typical habitat and neither are chimpanzees. These facts suggest that bonobos came first, giving rise to chimpanzees, which adapted to different habitats and thereby paved the way for man to evolve millions of years later. The bonobo, meanwhile, was left behind in the forest that it occupies today.

Another approach is adopted by John MacKinnon. He studied bonobos for a few weeks in 1975 and concluded that the reverse situation was more likely – that bonobos had actually evolved from chimpanzees. His argument is a twofold one, based on size and distribution.

Firstly, he points out that the adult bonobo is so similar to the immature chimpanzee in respect of body proportion, shape of skull and white tail tuft that it may have resulted from a special biological process known as neotony. By this process the immature stage of an individual progressing to adulthood becomes sexually mature while retaining its juvenile characteristics, thereby missing out a period of development and hastening the production of new young. Where environmental factors favor the process, natural selection can operate quickly to produce a new species, the individuals of which attain an adult stage closely resembling the juvenile form of their ancestor – hence the bonobo's small size and its retention in adult life of some of the characteristics of the juvenile chimpanzee.

Secondly, MacKinnon suggests that the very distribution of the bonobo could be direct evidence for its being a recent offshoot of the chimpanzee. This has something to do with the Zaire River. MacKinnon points out that many of the primates to the north have failed to cross this formidable barrier or are represented by different forms to the south.

Such a situation could have arisen after the enlargement of the river system which resulted in different conditions to the north and to the south. The two populations would then be subjected to different evolutionary forces leading to the speciation and subspeciation visible today.

Alternatively some of the northern apes could have spread around the headwaters of the Congo basin to invade the area on the south bank. Such a move would mean that the bonobo – or chimpanzee as it then was – once lived in more open woodland and has only recently invaded the rainforest. MacKinnon rests much of his case upon the lack of a harsh call. Normally in forest areas primates develop a system of calls by which group cohesion can be maintained. The bonobos apparently have none. Why? MacKinnon says that it is because the bonobo has only recently come to live in rainforest. Yet group maintenance must be one of the more critical features of life. Why have selective pressures not evolved contact calls? Perhaps – as both MacKinnon and the Badrians point out – it is because they have an alternative strategy. Bonobos will often descend to the ground and make a silent escape on all fours when disturbed in the trees. Since they have established terrestrial escape routes, it is possible that they know how to meet up again at some distant and safer spot in the forest. By travelling in small parties each individual might know precisely the whereabouts of the remaining members of its group and the need to evolve piercing cries which also serve to attract predators to the area could be overcome.

The common chimpanzee, *Pan troglodytes*, was not worked on seriously until well into the 20th century when Dr Henry Nissen spent two and a half months during 1931 in the forests of French Guinea. Unfortunately his study was too short to provide an 'in depth' appraisal of chimpanzee life. It was a useful beginning, but the real awakening force in chimpanzee study was the late Dr Louis Leakey. He and his wife Mary are today renowned

A mixed group of adult and young chimpanzees picks a single-file trail through savannah woodland, using the quadrupedal knuckle gait that they favor when moving in a relaxed fashion.

for their discoveries of early man fossils in Olduvai Gorge, Tanzania, but for more than 20 years before that Louis Leakey had foreseen the importance of a lengthy field study of chimpanzees. The remains of early man had often been found by a permanent water source, and a wild population of chimpanzees on the eastern shore of Lake Tanganyika held a special place in Leakey's mind. What he desperately needed was for somebody to go and work on them, to understand their lives and, he hoped, to shed new light on the daily lives of the humans that once lived in similar places. Leakey waited patiently for many years. One day at the end of the 1950s he was visited by a young English girl as he worked in the Coryndon Museum in Nairobi, Kenya. Her name was Jane Goodall and her story, now world famous, is a very remarkable one.

She was clearly destined to be a field worker. Among her earliest life-recollections is that of hiding away in a hen-house to see how a chicken laid an egg. After a triumphant watch, she emerged to find that her household was so worried about her disappearance that they had even contacted the local police. A few years later, at the age of eight, she apparently decided that she would one day work with animals in Africa – the 'country' of every child's romantic dreams, teeming with all the fascinating and frightening creatures of circuses and zoos.

Jane Goodall's dream strengthened to such an extent that later, when an old schoolfriend invited her to spend some time on her parents' farm in Kenya, she resigned from her job immediately and set about earning enough for her airfare by acting as a waitress in Bournemouth.

When she arrived in East Africa her love for the animals led her, perhaps inevitably, to Leakey's door. He employed her as his assistant and secretary and slowly introduced her to the idea of undertaking a vast chimpanzee research project at the Gombe Stream Chimpanzee Reserve (now called the Gombe National Park). She was completely unqualified academically but Leakey was not the sort of person to be influenced by that. He sensed her sympathy and her dedication and was happy that he had found the person he had sought for so many years. It is a credit to his imagination that such an opportunity came her way: it is a credit to her determination that she was prepared to go through with it all, especially in view of the fact that many people thought that she would last for no more than a few weeks.

In 1960 she began the project which was to keep her occupied for more than a decade. The influential Leakey arranged financial support, largely from the National Geographic Society, and helped to smooth the bureaucratic pathway that would allow a single girl to study man's closest living relative in the heart of Africa, with the Congo then in the throes of civil war no more than 25 miles away across the Western Rift Valley. (This was also Livingstone country, by the way, and it was on his *soko*, *Pan troglodytes schweinfurthii*, that she was now going to focus her existence.)

Her task was not an easy one. There were language barriers, tribal customs and local conditions to be absorbed. Most important of all she found she could get nowhere near the objects of her study. The chimpanzees were very wild and the terrain was such that her first sighting was by looking across a valley and onto the exposed canopy of a fruiting msulula tree, its small orange and red fruits occasionally obscured as a hairy black arm reached out to pick them, its boughs vibrating under the movements of visitors who remained mostly hidden from view.

After six months of acclimatization, she was rewarded with a major breakthrough. It came when her feelings of achieving success of even a moderate nature were running at their lowest ebb. She was tired, frustrated and almost at the end of another day's fruitless waiting for chimpanzees to accept her presence. The scream of a youngster attracted her attention and with it came the faint hope that she might catch sight of the evening nest-building in the forest. Silence. Then she saw four chimpanzees. A ten minute trek brought her to a spot from which she might watch them without precipitating their usual panic-stricken retreat into obscurity. Making her way cautiously to the cover of a large fig tree she peered round its age-twisted trunk to find that once again her presence had been long-detected and that she was apparently alone. She did not realize immediately that she was being closely watched, that there, not more than 20 yards from her, two male chimpanzees she had met before were sitting quietly with their eyes fixed upon her. Then a female and a youngster revealed themselves in the background and sat calmly in the lower branches of a tree. The two males groomed themselves and then one of them stood up and looked toward her. Her shadow, cast long by the setting sun, fell across his body. Later she saw a special significance in the moment. She called her book '*In the Shadow of Man*' and dedicated it in part to the memory of this male who she had already named David Greybeard.

It was in fact the same David Greybeard who provided Jane Goodall with two vital aspects of chimpanzee behavior. Both were revolutionary, occurred early on in her study and were largely instrumental in sparking off international interest in her work. From such an interest came the funds to keep her occupied for years.

One day before the end of her trial period at Gombe, she climbed to the top of a peak overlooking the study area and noticed a group of chimpanzees feeding in a tree slightly below her. David Greybeard was among them and he was holding a pinkish object in his hand. A female and a youngster watched him closely and occasionally offered an expectant hand as he pulled at the object with his teeth. Jane Goodall soon realized that the focus of their attention was meat. At one point David Greybeard dropped a piece to the ground and the youngster descended immediately. As he reached the ground, the undergrowth close by erupted and a large female pig charged out snorting violently, then pacing to and fro in a state of high tension. She had three piglets with her. When Jane Goodall approached to get a better view, she saw that the chimps were feeding on the remains of a young pig.

It was an exciting discovery. She obviously did not know at the time whether the chimpanzees had actually killed the piglet themselves but the significance of their actually eating it was important enough. Gorillas and orangs are strictly vegetarian, perhaps supplementing their diet with insects from time to time. Whether this is intentional or a consequence of vegetarianism – considering the strong association between plants and insects – cannot be satisfactorily deduced from an examination of animal droppings alone. It had previously been believed that chimps too were vegetarian. But here were apes at Gombe Stream actually eating flesh through intent.

Although meat-eating is a habit that had not been previously acknowledged for chimpanzees in the wild, there is an interesting early reference in *Nature* in 1889 to them doing so in captivity. Apparently a chimpanzee named Sally in London Zoo would kill and eat pigeons and sparrows that had the misfortune to venture into her cage. So distinctive was this habit that it was considered an important diagnostic feature. Young Sally could

This unusual picture captures a rare event in chimpanzees' social life – the communal hunting of bush pigs, which can be seen fleeing to the left across the rough track. Until the late 1960s it had been assumed that chimpanzees were vegetarian. Only with the publication of Jane Goodall's observations of meat-eating in 1971 did it become widely known that chimps are omnivorous, given the opportunity.

not be properly identified, partially on the grounds that the 'ordinary' chimpanzees never killed birds.

It has since been realized that chimpanzees do regularly kill and eat bush pigs, monkeys and even small antelopes. It is a habit that has been acquired mainly by the savannah dwelling chimpanzees, probably because the continuous lushness of the forest is missing from their diet.

It is worth stressing that man's own evolution almost certainly involved emerging from forest onto savannahs, a move that would also have necessitated a significant and lasting change in diet – from herbivore to omnivore.

Just a week or two after this revolutionary observation, David Greybeard furnished Jane Goodall with another startling piece of information. It was simply that chimpanzees *do* use natural objects as tools. Traditionally the use of tools was regarded as a specifically human activity; therefore, it was argued, apes do not use tools. True, the habit had been informally reported by people in West Africa where one chimpanzee had been seen to break open the kernels of palm nuts with a rock and a small group had been seen pushing twigs into a wild bees nest to get to the sweet, sticky honey. But she now saw it for herself. David Greybeard was sitting alone by a termite mound into which he was carefully poking a long piece of grass. He with-

Traditionally, tool-using was considered by anthropologists an activity unique to humans. Now it is established – again by Jane Goodall – that chimpanzees are tool users. They use rocks to split open nuts; they use sticks to probe for honey in bees' nests and, as these two pictures show, they use bits of grass to 'fish' for termites and ants. Above, a mother teaches her baby how to choose a suitable piece of grass and gives a lesson on fishing (right). The implications for man are startling.
As Jane Goodall has written, 'My early observations of their primitive tool-making abilities convinced a number of scientists that it was necessary to redefine man in a more complex manner than before.'

drew it after a few moments, raised it to his mouth and plucked a small object from the end. It was a termite. He fed for at least an hour before ambling off and allowing Jane Goodall to move in and examine the feeding site. A swarm of termites was busily sealing the enlarged hole made by the intruding chimpanzee and when she introduced a grass stem into it herself, she felt an unmistakable tug as it was attacked below ground. When she removed it, several worker and soldier termites were stuck fast to it, locked securely in place by their strong jaws.

After eight days of vigilance in a rough hide near the termite mound, Jane Goodall was again rewarded. David Greybeard returned with another male, Goliath, and the pair of them began feeding as she had witnessed before. But this time they also did something different. Several times they picked leafy twigs, stripped them of their greenery and then used the bare stems for feeding. Here were apes not only *using* tools, but *making* them – rather crude ones, perhaps, but still another example of behavior that for so long had been considered the sole prerogative of man. It opened up the crucial issue of intelligence and how far apes could adapt their behavior in a

An adolescent female watches curiously as her brother prepares a fruit and then raises it to his mouth. Even such a simple sequence reveals the sophistication of chimpanzee relationships: she, as the little sister, is aware of the dangers of trying to grab the fruit and he, knowing that she knows her place, tolerates her attention, which in a stranger would be an unthinkable intrusion.

way previously thought to be essentially human.

Chimpanzee social life differs enormously from that of other apes. Gorillas live in more or less fixed groups and orangs are largely solitary or at least found in widely spaced social units. With chimpanzees, group sizes can vary appreciably from day to day as individuals make their way around a large area, spending some of their time with one group and then leaving abruptly to join another. There can, therefore, be no strict hierarchy within each group, for such a strategy would not tolerate the temporary presence of 'outsiders.' Perhaps a less rigid diet found over a wide area permits the free ranging of individuals, and social life has evolved accordingly to prevent the formation of stable and exclusive units. As many as 80 individuals may form a chimpanzee community. In forests where food is abundant and fairly evenly distributed they may roam over an area of up to ten square miles, but in savannah woodlands where they move around much more the same number may occupy at least 30 square miles.

Within each chimpanzee group there are, however, dominant and subordinate individuals. This state of affairs concerns males more than females

Male chimpanzees establish their position in the social hierarchy by pant-hooting (left) and then by charging (right) in a display of aggression that is more noise than substance. Male adolescents, in particular, assert themselves by screaming and throwing objects during a charge. Such displays seldom lead to outright violence.

because they usually remain within the confines of the group into which they are born while females may easily leave and join another group to breed. Such a system overcomes any possible problems of inbreeding and serves to maintain a wide-ranging community of healthy individuals. So within a group it is the nature of males to jockey for power. One of them dominates the rest and enjoys a few feeding and social privileges in accordance with his status.

But life at the top is usually short-lived, and as a result most of the resident males, many of whom will live between 30 and 40 years, enjoy a spell of supremacy. Ranking is achieved and maintained through intensity of dominance displays. An individual will sway back and forth on his haunches, slowly erecting the hairs on his body to increase his apparent size. At the same time he begins to hoot softly and as these sounds increase in intensity so he becomes more agitated and excited. At peak size and sound he launches himself towards his high-ranking 'victim' thrashing around, grabbing sticks and branches and unleashing a seemingly unprovoked fury upon grasses and plants all around him. If he has impressed his adversaries to the extent that they flee in terror from his epileptic advances then they will inform him of his victory by returning submissively to his side, grooming him and giving him every physical and emotional reassurance that his antics have won the day. Of course he may not succeed in his ambitions and will have to retreat himself to wait in the background until the moment seems right once more.

Once these short, maniacal bouts are over and the status of all concerned is agreed upon, a friendly calm descends and all continue as though nothing out of the ordinary has taken place. Evolution has worked matters in such a way that the quarrels are short, to the point and immediately forgotten so as not to disrupt chimpanzee life for more than the minimum amount of time required to resolve leadership problems.

Jane Goodall observed at Gombe a fascinating situation in which one of her study males elevated himself in status through the use of resonant man-made objects. These were empty four-gallon paraffin cans which when soundly banged together produced a noise more terrifying than that produced by any chimpanzee. The individual in question was called Mike and he was apparently very near the bottom of the dominance hierarchy. In such a position he had to feed after the others (especially at an artificial supply of bananas) and was all too often subjected to spiteful reminders of his inferiority.

Toward the end of 1963 Jane Goodall left Gombe for four months, but she returned to find a completely different Mike whose ride to power had been documented by students in her absence. She soon saw for herself the way in which Mike's status had been improved, although he still had some way to go before he became the highest ranking male in the group. He had, quite simply, learned to incorporate the empty cans into his displays and as the producer of such frightening sounds was enjoying the increased respect of the chimpanzees around him.

Goodall soon saw for herself what had been taking place during her absence from camp life. One day five superior males, including the number one, Goliath, were quietly grooming each other while Mike sat alone some yards apart from them. He got up, walked over to a tent and grabbed hold of an empty paraffin can. Then he took hold of a second can and walked quietly back to his resting place, sat down and stared toward the unsuspecting group which paid him little attention. Suddenly he began to rock gently from side to side, calling softly in preparation for an attack. The tempo of his display increased and his hair became erect. Then he was on his feet and charging, hooting loudly, holding the cans out in front of his body and crashing them together. The combination was too much and the select party of males ahead of him scattered out of his way as he con-

In chimpanzee society, the patterns of dominance shift continuously, and gestures of reassurance and submission are essential ways of defusing tension. At left, a dominant male reassures another male by touching his genitals, and below a submissive male, having presented his rump, is mounted by a dominant male in mock copulation.

tinued his charge down a track, out of both sight and sound. But he was back again before the perplexed group could reassemble properly. On his return charge Mike went straight for the top – at Goliath himself, who also beat a hasty retreat. In defiance Mike sat and glared, breathing heavily and waiting for a response. One by one the males, except Goliath, returned to him, crouching submissively, touching and grooming him. Mike had

obviously taken second place to Goliath who must have sensed that he, too, might soon have to submit to the new wave of terror.

The transition as Mike took over power was a gradual one. Goliath could not hang on for much longer, for Mike's displays were both frequent and effective. There came a time when even the human occupants of the camp feared for their safety because Mike soon realized that if he let go of one of his cans while he was swinging it around, the effect could be even more devastating. So his ritual success-objects were kept well out of his way and, although he may have felt rather vulnerable, his top-ranking status was by this time assured.

Within this arena of male dominance are females who once every five weeks or so are sexually receptive. Their state is heralded by a prominent swelling and pinkening of their rear end (ano-genital region). It seems that a large part of the sexual activity that follows these swellings is undergone more for social than for reproductive reasons, for chimpanzees are highly promiscuous. Only rarely does a female conceive under such circumstances. It is interesting to wonder what mechanism(s) – either physical or hormonal – might be responsible for this prevention. What is more certain is that a male and a female may suddenly be conspicuous by their absence from a group, to return a week or two later when the female is often pregnant. It may be that repeated copulation with the same male, with the female submitting in an appropriate position to facilitate the deeper penetration of his sperm, is required before she can conceive. It may be something that pairs of animals can achieve more satisfactorily on their own, undisturbed and in private. It may also be that social sex is an important ingredient of

Like human children, chimpanzee children are dependent on their mothers for several years. Most youngsters continue to suckle and sleep with their mothers for over four years.

A male prepares to mount a presenting female. Females menstruate about every 35 days. Between menstrual periods, when they are at their most fertile, females develop a pale pink swelling around the genital area, which they display as a sexual signal to males. Typically, copulation lasts no more than about 15 seconds.

This drawing of a keeper and his chimpanzee charge in the London Zoo in 1874 shows the poverty of the environments to which captive chimps were – and all too often still are – subjected.

community maintenance and that reproductive sex has evolved on a more complicated basis. Both aspects are vital to chimpanzees' welfare. It is tempting once more to compare these findings with present-day human society in the West, where non-reproductive sex has also assumed an important social role.

If the female chimpanzee has conceived she will be pregnant for between 200 and 260 days. Her youngster will be born into an essentially female dominated group which may easily contain an older sister who will watch and learn the art of motherhood first-hand. She will see the newborn infant cling helplessly to its mother, held close to her breast where it can feed at frequent intervals. Because chimpanzees produce young only once in every four or five years, births are rare events and other females and even males cluster round and show a great deal of interest, touching and caressing the new arrival. When its mother moves around, the baby hangs on to her belly, its tiny hands gripping her hair.

The ensuing months see a development in both strength and size and the change is marked by a switch from riding below to riding on the mother's back – a position that is probably much more comfortable for both individuals. When she rests, the baby chimp can slide off her back and explore the surrounding area for itself, but always under her watchful gaze. After a year or so of gradual development, the youngster is playing with any other juveniles that may be available and willing to join in. At the same time it will be experimenting with any food it can find, although it is still largely dependent upon its mother for nourishment.

The upbringing is a lengthy affair, as befits any animal with a long life expectancy and low reproductive rate, for – as in humans – a great deal is invested by parents (especially the females) in each offspring and its chances of survival must be maximized. The play which characterizes so much of this childhood must be vital to familiarize the youngster not only with its environment but also with the older individuals into whose society it is being projected. Among themselves, the young soon learn who is strongest, who responds to attack or to bluff and whose mother is held in high esteem within the female group. In short it is the intertwining and

The mother-child bond is as strong in captivity as it is in the wild. Here a mother tickles her child playfully, instinctively creating the strong emotional bond between the two that is necessary to ensure the youngster's survival and the acquisition of maturity.

An adolescent female greets an adolescent male with a kiss. After a separation, the behavior of two chimpanzees who know each other often looks amazingly like that shown by humans. They may bow or crouch, hold hands, kiss, embrace, touch or pat each other – all gestures of affection and reassurance.

understanding of characters which themselves take years to develop.

The young male is physically mature at about seven or eight years of age, although he still has a considerable amount of growing ahead of him before he reaches his full 100-pound weight. He is also low-placed on the dominance scale and while he may show intimidating behavior towards adult females, he must still be very careful not to antagonize the older males around him. His relationship with his mother remains a critical factor, for he may return to her at any time, even well into his early teens, for protection. There may come a time later when, as a fully independent adult, he will return to her side when she is in any difficulty. It seems to be a loyal bond, one which is loosened but not broken after many years. The young male may also spend quite a lot of time on his own. He does not yet fit into the dominance hierarchy of other males (if he does join them he may be badly treated) and neither can he stay with his mother, especially if she is tending another youngster. His bouts of enforced loneliness probably serve to sharpen the edge of his acquired survival techniques although he may just follow other chimpanzees around while maintaining a healthy distance. But by his fifteenth year or so, he is gradually working his way into the male oriented groups, there to enjoy their company, take repeated abuse from them and wait for the day when he can work his way up the scale to supremacy.

The young female also becomes physically mature at about seven or eight years of age. Although her mother becomes increasingly less tolerant of her presence, she is permitted to follow her around and will be allowed to help with the nursing of a newborn youngster. She is probably more attracted to juveniles than she ever was as one of them herself, and this is probably a critical period regarding her future as a successful mother. By the time she is about nine she begins to exhibit the full sexual swellings and the males become interested in her state. They (one and as many more that are around) puff themselves up and demand her attention by shaking branches as they approach. She can do little to stop the onslaught as she is pursued and subjected to increasingly vigorous courtship. There is something terribly inevitable about the whole procedure as she finally crouches down and submits to the onslaught, screaming.

But not all females share this innate terror of their first sexual encounter. Indeed Jane Goodall noticed one who was possibly the chimpanzee equivalent of a nymphomaniac. Her name was Fifi and she took a deep interest in copulating couples long before she was herself ready to be noticed. When the 'day' finally arrived, she responded immediately to the first move of an eligible male and even went rushing over to others to make sure that they had noticed her newly-acquired status. She was most surprised when her swellings went down and the males were no longer interested in her, no matter how hard she tried to rouse them.

The female is subjected to these periodic sexual encounters for at least two years before she finally gives birth to her own offspring. It seems that some physiological or hormonal mechanism operates to prevent conception during socialized sexual behavior, when the young female is still not ready for the responsibilities of motherhood. She is not fully grown yet and neither is she at all secure socially. Perhaps it is only when she had acquired her full status within the group that she becomes eligible for one of the secret 'honeymoons' that, about eight months later, will result in her bringing her first newborn chimpanzee into the world, there to develop and to learn for itself all the intricacies of a life that is in so many ways a shadow of our own.

JANE GOODALL LOOKS INTO 'THE SHADOW OF MAN'

'I had been there some fifteen minutes when a slight movement on the bare burnt slope just beyond a narrow ravine caught my eye. I looked round and saw three chimps standing there staring at me. I expected them to flee, for they were no farther than eighty yards away, but after a moment they moved on again, quite calmly, and were soon lost to sight in some thicker vegetation. Had I been correct, after all, in my assumption that they would be less afraid of one person, completely alone? For, even when I had left my African companions behind and approached a group on my own, the chimps had undoubtedly been fully aware of what was going on.

I remained on my peak and later on in the morning a group of chimps, with much screaming and barking and pant-hooting, careered down the opposite mountain slope and began feeding in some fig trees that grew thickly along the stream banks in the valley below me. They had only been there about twenty minutes when another procession of chimps crossed the bare slope where, earlier, I had seen the three. This group also saw me – for I was very conspicuous on the rocky peak. But, although they all stopped and stared and then hastened their steps slightly as they moved on again, the chimpanzees did not run in panic. Presently, with violent swaying of branches and wild calling, this group joined the chimpanzees already feeding on figs. After a while they all settled down to feed quietly together, and when they finally climbed down from the trees they moved off in one big group. For part of the way, as they walked up the valley, I could see them following each other in a long, orderly line. Two small infants were perched, like jockeys, on their mothers' backs. I even saw them pause to drink, each one for about a minute, before leaping across the stream.

It was by far the best day I had had since my arrival at Gombe, and when I got back to camp that evening I was exhilarated, if exhausted. Vanne, who had been far more ill than I and who was still in bed, was much cheered by my excitement.

That day, in fact, marked the turning-point in my study. The fig trees grow all along the lower reaches of the stream and, that year, the crop in our valley was plentiful and lasted for eight weeks. Every day I returned to my peak, and every day chimpanzees fed on the figs below. They came in large groups and small groups, singly and in pairs. Regularly they passed me, either moving along the original route across the open slope just above me, or along one or other of the trails crossing the grassy ridge below me. And, because I always looked the same, wearing similar dull-coloured clothes, and because I never tried to follow them or harass them in any way, the shy chimpanzees began to realise, at long last, that I was not, after all, so horrific and terrifying. Piece by piece, I began to form my first somewhat crude picture of chimpanzee life.

The impression that I had gained when I watched the chimps at the msulula tree of temporary, constantly changing associations of individuals within the community was substantiated. Most often I saw small groups of four to eight moving about together. Sometimes I saw one or two chimpanzees leave such a group and wander off on their own or join up with a different association. On other occasions I watched two or three small groups joining up to form a larger one.

Often, as one group crossed the grassy ridge separating the Kasakela Valley from the fig trees in the home valley, the male chimpanzees of the party would break into a run, sometimes moving in an upright position, sometimes dragging a fallen branch, sometimes stamping or slapping on the hard earth. These charging displays were always accompanied by loud pant-hoots, and afterwards the chimpanzee often swung up into a tree overlooking the valley he was about to enter, and sat quietly peering down and obviously listening for a response from below. If there were chimps feeding in the fig trees they nearly always hooted back, as though in answer. Then the new arrivals hurried down the steep slope and with more calling and screaming, the two groups met up in the fig trees. When groups of females and youngsters, with no males present, joined other feeding chimpanzees there was usually none of the excitement; the newcomers merely climbed up into the trees, greeted some of those already there, and began to stuff themselves with figs.

Whilst many details of their social behavior were hidden from me by the foliage, I did get occasional fascinating glimpses. I saw one female, newly arrived in a group, hurry up to a big male and hold her hand towards him. Almost regally he reached out, clasped her hand in his, drew it towards him and kissed it with his lips, I saw two adult males embrace each other in greeting. I saw youngsters having wild games through the tree-tops, chasing around after each other or jumping again and again, one after the other, from a branch to a springy bough below. I watched small infants dangling happily by themselves for minutes on end, patting at their toes with one hand, rotating gently from side to side. Once two tiny infants pulled on opposite ends of a twig in a gentle tug-of-war. Often, during the heat of midday, or after a long spell of feeding, I saw two or more adults grooming each other, intently looking through the hair of their companions.

At that time of year the chimps usually went to bed late, making their nests when it was too dark to see properly through binoculars, but sometimes they nested earlier so that I could watch them from the Peak. I found that every individual, except for infants who slept with their mothers, made his own nest each night. Usually this took about three minutes: the chimp chose a firm foundation, such as upright fork or crotch, or two horizontal branches. Then he reached out and bent over smaller branches on to this foundation, keeping each one in place with his feet. Finally he tucked in the little leafy twigs growing round the rim of his nest and then lay down. Quite often a chimp sat up after a few minutes and picked a handful of leafy twigs which he put under his head or some other part of his body before settling down again for the night. One young female I watched went on and on bending down branches until she had constructed a huge mound of greenery on which she finally curled up.

I climbed up into some of the nests after the chimpanzees had left them – most of them were built in trees that, for me, were almost impossible to climb. I found that there was quite complicated interweaving of the branches in some of them. I found, too, that the nests were never fouled with dung – and later, when I was able to get closer to the chimps, I saw how they were always careful to defecate and urinate over the edge of their nests, even in the middle of the night. . . .

The light rains of the Chimpanzees' Spring gave place to the long rains. Showers became deluges which sometimes lasted with unabated fury for two hours or more. One of those wild storms, which occurred a week or so after the change of seasons, I shall long remember. For two hours I had been watching a group of chimpanzees feeding in a huge fig tree. It had been grey and overcast all morning, with thunder growling in the distance.

At about noon the first heavy drops of rain began to fall. The chimpanzees climbed out of the tree and, one after the other, plodded up the steep grassy slope towards the open ridge at the top. There were seven adult males in the group, including Goliath

Jane Goodall has gained the confidence of her subjects to such an extent that they will take bananas from her and even play with her clothes.

and David Greybeard, several females, and a few youngsters. As they reached the ridge the chimpanzees paused. At that moment the storm broke. The rain was torrential and the sudden clap of thunder, right overhead, made me jump. As if this was a signal, one of the big males stood upright and, as he swayed and swaggered rhythmically from foot to foot, I could just hear the rising crescendo of his pant-hoots above the beating of the rain. Then he charged off, flat-out down the slope towards the trees he had just left. He ran some thirty yards and then, swinging round the trunk of a small tree to break his headlong rush, leapt into the low branches and sat motionless.

Almost at once two other males charged after him. One of them broke off a branch from a tree as he ran and brandished it in the air before hurling it ahead of him. The other, as he reached the end of his run, stood upright and rhythmically swayed the branches of a tree back and forth, before seizing a huge bough and dragging it farther down the slope. A fourth male, as he too charged, leapt into a tree and, almost without breaking his speed, tore off a large branch, leapt with it to the ground and continued down the slope. As the last two males called and charged down, so the one who had started the whole performance climbed from his tree and began plodding up the slope again. The others, who had also climbed into trees near the bottom of the slope, followed suit. And then, when they reached the ridge, they started charging down all over again, one after the other, with equal vigor.

The females and youngsters had climbed into trees near the top of the rise as soon as the displays had begun, and there they remained watching throughout the whole performance. As the males charged down and plodded back up, so the rain fell harder and harder, jagged forks or brilliant flares of lightning lit the leaden sky and the crashing of the thunder seemed to shake the very mountains.

My enthusiasm was not merely scientific as I watched, enthralled, from my grandstand seat on the opposite side of the narrow ravine, sheltering under a sheet of polythene. In fact, it was raining and blowing far too hard for me to get at my notebook or use my binoculars: I could only watch, and marvel at the magnificence of those splendid creatures. With a display of strength and vigor such as this, primitive man himself might have challenged the elements.

Twenty minutes from the start of the performance the last of the males plodded back up the slope for the last time. The females and youngsters climbed down from their trees and the whole group moved over the crest of the ridge. One male paused and, with his hand on a tree-trunk, looked back – the actor taking his final curtain. Then he too vanished over the ridge.

I continued to sit for a while, staring almost in disbelief at the white scars on the tree-trunks and the discarded branches on the grass – all that remained, in that rain-lashed landscape, to prove that the wild "rain dance" had taken place at all.

I should have been even more amazed had I known then that I should only see such a display twice more in the next ten years. Often, indeed, male chimpanzees react to the start of heavy rain by performing a rain dance, but this is usually an individual affair. Yet I have only to close my eyes to see again, in vivid detail, that first spectacular performance. . . .

To many scientists interested in human behavior and evolution, one very significant aspect of chimpanzee behavior lies in the close similarity of many of their communicatory gestures and postures to those of man himself. Not only are the actual positions and movements similar to our own, but also the contexts in which they often occur.

When a chimpanzee is suddenly frightened he frequently reaches to touch or embrace a chimpanzee close by, rather as a girl, watching a horror film, may seize her companion's hand. Both chimpanzees and humans seem reassured, in stressful situations, by physical contact with another individual. Once David Greybeard caught sight of his reflection in a mirror; terrified, he seized Fifi, then only three years old. Even such contact with a very small chimp seemed to reassure him; gradually he relaxed and the grin of fear left his face. Humans, indeed, may sometimes feel reassured by holding or stroking a dog, or some other pet, in moments of crisis.

This comfort which the chimpanzee and human alike appear to derive from physical contact with another, probably originates during the years of infancy when, for so long, the touch of the mother, or the contact with her body, serves to calm the frights and soothe the anxieties of both ape and human infants. So, when the child grows older and his mother is not always close by, he seeks the next best thing – close physical contact with another individual. If his mother is around, however, he may deliberately pick her out as his comforter. Once when Figan was about eight years old, he was threatened by Mike. He screamed loudly and hurried past six or seven other chimps nearby until he reached Flo; then he held his hand towards her and she held it with hers. Calmed, Figan stopped screaming almost at once. Young humans too continue to unburden their hearts to their mothers long after the days of childhood have passed; provided, of course, that an affectionate relationship exists between them.

There are some chimps who, far more than others, constantly seem to try to ingratiate themselves with their superiors. Melissa, for one, particularly when she was young, used to hurry up and lay her hand on the back or head of an adult male almost every time one passed anywhere near her. If he turned towards her, she often drew her lips back into a submissive grin as well. Presumably Melissa, like the other chimps who constantly seem to try to ingratiate themselves in this way, is simply ill at ease in the presence of a social superior so that she constantly seeks reassurance through physical contact. If the dominant individual touches her in return, so much the better.

There are, of course, many human Melissas: the sort of people who, when trying to be extra friendly, reach out to touch the person concerned and smile very frequently and attentively. Usually they are, for some reason or other, people who are unsure of themselves and slightly ill at ease in social contexts. And what about the smiling? There is much controversy as to how the human smile has evolved. But it seems fairly certain that we have two rather different kinds of smile, even if, a long time ago, they derived from the same facial gesture. We smile when we are amused and we smile when we are slightly nervous, on edge, apprehensive. Some people, when they are nervous at an interview, smile in this way at almost everything that is said to them. And this is the sort of smile that can probably be closely correlated with the grin of the submissive or frightened chimpanzee.

When chimpanzees are overjoyed by the sight of a large pile of bananas they pat and kiss and embrace one another rather as two Frenchmen may embrace when they hear good news, or as a child may leap to hug his mother when told of a special treat. We all know those feelings of intense excitement or happiness which cause people to shout and leap around, or to burst into tears. It is not surprising that chimpanzees, if they feel anything akin to this, should seek to calm themselves by embracing their companions.

I have already described how a chimpanzee, after being threatened or attacked by a superior, may follow the aggressor, screaming and crouching to the ground or holding out his hand. He is, in fact, begging a reassuring touch from the other. Sometimes he will not relax until he has been touched or patted, kissed or embraced; Figan several times flew into a tantrum when such contact was withheld, hurling himself about on the ground, his screams cramping in his throat, until the aggressor finally calmed him with a touch. I have seen a human child behaving in the same sort of way, following his mother around the house after she has told him off, crying, holding her skirts, until finally she picked him up and kissed and cuddled him in forgiveness. A kiss or embrace or some other gesture of endearment is an almost inevitable outcome once a matrimonial dispute has been resolved, and the clasping of hands to denote renewal of friendship and mutual forgiveness after a quarrel occurs in many cultures.

It is if we begin to consider the moral issues at stake when one human begs forgiveness from another, or himself forgives, that we get into difficulties when trying to draw parallels between human and chimpanzee behavior. In chimpanzee society the principle involved when a subordinate seeks reassurance from a superior, or when a high-ranking individual calms another, is in no way concerned with the right or wrong of the aggressive act. A female who is attacked for no reason other than that she happens to be standing too close to a charging male is quite as likely to approach the male and beg a reassuring touch as is the female who is bowled over by a male as she attempts to take a fruit from his pile of bananas.

Again, whilst we may make a direct comparison between the effect, on anxious chimpanzee or human, of a touch or embrace of reassurance, the issue becomes complicated if we probe into the motivation which directs the gesture of the ape or the human who is doing the reassuring. For humans are capable of acting from purely unselfish motives; we can be genuinely sorry for someone and try to share in his troubles in an endeavour to offer comfort and solace. It is unlikely that a chimpanzee acts from feelings quite like these; I doubt whether even members of one family, united as they are by strong mutual affections, are ever motivated by pure altruism in their dealings one with another.

On the other hand, there may be parallels in some instances. Most of us have experienced sensations of extreme discomfort and unease in the presence of an abject, weeping person. We may feel compelled to try to calm him, not because we are sorry for him, in the altruistic sense, but because his behavior disturbs our own feeling of well-being. Perhaps the sight – and especially the sound – of a crouching, screaming subordinate similarly makes a chimpanzee uneasy; the most efficient way of changing the situation is for him to calm the other.

There is one more aspect to consider in relation to the whole concept of reassurance behaviour in chimpanzees, and that is the possible role played by social grooming in the evolution of the behavior. For the chimpanzee – and indeed for many other animals too – social grooming is the most peaceful, most relaxing, most friendly form of physical contact. Infant chimpanzees are never starved for physical contact for they spend much time close to their mothers. Then, as they get older, they spend more time away from their mothers and also more time playing with other youngsters; and play, typically, involves a good deal of physical contact. As the youngster matures he gradually plays less frequently; at the same time he spends longer and longer socially grooming, either with his mother and siblings or, as he gets older, with other adults. Sometimes a grooming session between mature individuals may last for two hours. The obvious need for social grooming was well demonstrated when old Mr McGregor, with his paralysed legs, dragged himself those sixty long yards to try and join a group of grooming males.

When a chimpanzee solicits grooming he usually approaches the selected partner and stands squarely in front of him, either facing him with slightly bowed head or facing away and thus presenting his rump. Is it possible, then, that submissive presenting of the rump, and submissive bowing and crouching, may have derived from the postures used to solicit grooming? That in the mists of the past the subordinate approached his superior, after he had been threatened, to beg for the reassuring, calming touch of grooming fingers? If so, then the response of the chimpanzee thus approached, the touch or the pat, may equally have been derived from the grooming pattern. Indeed, on some occasions a few brief grooming movements do occur when a dominant individual reaches out in response to the submissive posture of a subordinate. It is quite reasonable to suppose that such a response may have become ritualised over the centuries so that to-day the chimpanzee usually gives a mere token touch or pat in place of grooming his submissive companion.'

Goodall with Lulu, National Zoo, Washington.

5/THE PATH TO MAN

Charles Darwin predicted more than 100 years ago that *Homo sapiens* would be found to have evolved in Africa. It is now accepted that he was right. Fossil bones and stone artefacts found over the last decade in Tanzania, Kenya (at places like the desolate Koobi Fora site on the shores of Lake Turkana, left) and Ethiopia have extended human history back over five million years. It now seems certain that our pre-human ancestors shared their African homeland with creatures – the Australopithecines – to whom they were closely related, but yet who vanished into evolutionary oblivion. The details are still missing – and will almost certainly remain so for many years – but at least we now have a framework within which to understand our relationship to the great apes.

Although there is now overwhelming evidence to support the theory of evolution by natural selection, and of man and apes sharing a common ancestry, the details are still controversial – indeed the idea itself is controversial.

As long ago as 1758, Carl Linnaeus, the first serious student of classification, officially recognized the close relationship between humans, monkeys and apes. He devised the group name – or Order – Primates to encompass them all and to denote their high ranking in the organization of the animal kingdom. Linnaeus made a few mistakes (such as including bats in this Order), but his was the first scientific recognition that man, *Homo sapiens*, had strong similarities with other animals. He was not unduly concerned with the means by which the likenesses had been achieved, just that they should be recognized.

His was not an exactly revolutionary piece of scientific pigeon-holing, for any person could see how much more closely a chimpanzee – not that there were too many of them around then – resembled the human than did a cat or a dog. This had always drawn attention. A hundred years or so before the birth of Christ, Pliny had recognized the similarities between man and monkeys, even elevating some forms of the latter group to the status of 'wild men.' Similarity implies relationship. Linnaeus had in effect sown the first real seeds of doubt.

At the time it was generally held that species were fixed. The arrangement of the animal kingdom was seen as a precise hierarchy of increasingly complex forms, each created independently of the ones to either side of it – beginning with the humblest known and culminating with man who was the only one to have been made exactly in the image of God.

There were inconsistencies in the idea. By the 17th century a great number of fossil bones had been collected in Europe, particularly those of the ice-age mammoth, which presented certain problems. With no adequate dating mechanisms available, let alone any serious belief that they were at all necessary, it was easy to dismiss such relics as of recent origin. What was not quite so easy to explain away was their great size. Apparently there had been other species that no longer existed. (At one point the idea of a giant form of early man was suggested as the only explanation; a Frenchman named Henrion calculated that Adam and Eve may have attained the super-mammoth height of about 120 feet!)

The answer was found in the Flood, which, it was argued, destroyed both early man and a number of species not mentioned in the Bible. Linnaeus's work did not threaten this superficial reassuring theory. He, too, believed that all animals had been created by the hand of God and that man had been set apart from them in a special way.

His arch rival of the day was the Comte de Buffon, the celebrated French naturalist, who saw towards the end of the 18th century that nature was a succession of movements, a biological flow – as ancient Greek philosophers before him had seen. His ideas were seized upon by one of his students, Jean-Baptiste Pierre Antoine de Monet, Chevalier de Lamarck. Lamarck saw correctly that all species are subject to change. He even said that varying environmental conditions imposed themselves upon animals to such an extent that the animals gradually changed from one species to another. Perhaps his greatest understanding was that such changes were effected at a very slow rate and would manifest themselves after only a very great deal of time. It was a masterpiece of evolutionary thought, for it included man and even presented a way in which a four-legged animal could become a

Carl von Linné (or Linnaeus, the Latin version of his name by which he is generally known) recognized the need to classify living things according to their similarities. He did not suggest relationships based on common ancestry – he accepted the Biblical view of the origin of species – but his scheme was surprisingly accurate and remains the basis for the modern system of evolutionary classification.

two-legged one and of how an arboreal two-legged one could become a terrestrial two-legged one; there, down on the ground, it could develop the senses and the abilities which would enable it to dominate all other forms of animal life.

But as Director of the Jardin des Plantes in Paris Lamarck could not make such suggestions without proposing a process by which it might have happened. It was in such a proposition that his downfall lay and, rather sadly, his unfortunate reputation in the 20th century. He said that an animal would undergo such changes during its lifetime and that the bodily alteration would be passed on to any young that that individual produced. Thus favorable characteristics were to be inherited by offspring, which could themselves change during their lifespan. It was a bold effort, close in many ways, and yet wrong. His theory could not possibly explain the amount of variation between individual offspring.

Any good Lamarck may have done the cause of evolution was undone by Baron Georges Cuvier, the zoological king of the opening 30 years of the 19th century. While his contributions to paleontology were indeed immense, he categorically pronounced that man had nothing to do with the ancient animal life to which the fossil records so obviously testified. There could be no such thing as a fossil human, he declared, for man had come into existence suddenly and recently and was quite separate from the rest of the animal kingdom.

Yet there were many pieces of evidence that belied man's supposedly recent, divine origins. By the early 19th century a great wealth of evidence existed of past human activity: hand-axes, spearheads, arrowheads and all sorts of roughly hewn objects, some of stone, some bronze and some iron. All undoubtedly had been fashioned and employed by the hands of men. But which men? Was there not a progression? Could it be reconciled with the Biblical time-span – a span that had been precisely dated by Archbishop Ussher, who stated that the world was formed in 4004 BC on Saturday, 3 October at 8.00 pm.

One of the greatest steps toward a solution to the problem was taken in the emerging field of geology. The first person to acknowledge the possibility that the forces which had shaped the surface of the earth – its mountains, continents, rivers – was James Hutton. In the late 18th century he concluded that not only were these forces still in operation but that the world was also a good deal older than had been previously acknowledged – suggestions that found no general acceptance. Hutton died in 1797, the year that Charles Lyell was born, Hutton's intellectual heir. Lyell was particularly interested in some remains of early man's stone implements which had been found in northern France. He believed that they were genuinely ancient. His greatest work, *The Principles of Geology*, demonstrated how the face of the earth had changed over an immense period of time. And he also saw, although it took him many years to reconcile himself to the fact, that there was no reason why animal life should not have changed over a similar period.

By the end of the 18th and well into the 19th century, two distinct lines of approach had appeared. There were still those who saw all forms of life as having been created by the hand of God, and there was a now hardening core of evolutionists who thought in terms of gradual processes of complexity.

One of these latter men was Erasmus Darwin, an early evolutionist who suggested that all creatures may have been derived from a common ancestry and that, as a consequence of this, species must be able to change from one to the other. As to the mechanisms involved, he suggested factors outside the bodies of individuals which might be of significance. He saw competition as an important ingredient and maintained that overpopulation was a vital factor in intensifying competition. He also saw the importance of the environment as a tool for fashioning species.

It fell to Erasmus's grandson, Charles, to formulate a framework that made the fact of evolution undeniable. Moreover he knew as well as anyone that man could not help but be included in his results, although he avoided the implications of this aspect of the theory until later in life. When Darwin set off on his famous journey in the *Beagle* under the captaincy of Robert Fitzroy in 1831, he was simply in quest of knowledge. He had read all the available literature and although he had graduated from Cambridge University with a degree in Theology and no formal scientific qualifications, his mind was well prepared to assess the evidence he found.

His major proof of evolution rested on specimens collected on the Galapagos Islands off the coast of Ecuador. There he had obtained a good series of finches from neighboring islands and although the birds were superficially similar, they differed greatly in respect of the size of their beaks. Darwin concluded that at some stage they had all had a common ancestor whose offspring had changed as the population spread out. The change did not have to be drastic, just enough to allow different groups (species) to live side by side, even on the same island, without competing for food. Some of them were thick-billed and adapted for seed-eating while others were thin-billed and adapted for insect-eating. Yet another had evolved the trick of using a cactus spine, in a way reminiscent of the termite-fishing antics of the chimpanzee, to obtain insects otherwise unavailable to it. Other animals on the islands were adapted in their own ways.

When he returned to England in 1836 Darwin set to work, committed to a lifetime of attention to detail and experimentation. It took him 20 years to formulate his conclusion, and even then he was pushed into publication. Another man, Alfred Russel Wallace, hit upon the same solution quite independently of Darwin and at exactly the same time. Wallace was committed to years of work in Southeast Asia during the mid-19th century and, like Darwin, he stockpiled a mass of specimens which gave him the vital clues to an understanding of the formation of species. It was Wallace who sent to Darwin in 1858 a letter putting forward the theory he had formulated – that species evolved by the steady selection by nature of those fittest to survive and breed. Darwin gave a joint paper from the two men to the Linnaean Society in the same year. When he published *On the Origin of Species by Means of Natural Selection* in November 1859, it sold out.

The theory of evolution by natural selection expressed by Darwin (seen top at the age of 40 when he was still working on *The Origin of Species*) applied to man as well as all other animals – a fact that lay in part behind his reluctance to publish and behind the passionate antagonism to his ideas, exemplified by the cartoon above dismissing Darwin as an ape.

Both Darwin and Wallace had seen that changing environments could cause changes among species, not during their lifetime as Lamarck had suggested, but from generation to generation. The clue, therefore, was in reproduction and the amount of individual variation in the young produced. These variable young would then, by virtue of their birth alone, be 'offered' to the environment. The best varieties for the conditions prevailing would survive (or at least breed most successfully) and a high proportion of their characteristics would prevail in the next generation. The vital point is that environments themselves undergo changes and animals must evolve new forms that are always well adapted to live and reproduce. Thus one species can in time become different – ultimately so different that it becomes a new species that will not breed with its original form.

The consequence of all this has been, over a few thousand million years, the evolution of the wildlife spectrum, plants included, that can be found in the world today. The inescapable conclusion, the one that caused Darwin so much concern and the one which caused so much controversy within the establishment, is that man is an integral part of this biological process; that man is an animal, a product of countless past variations selected by the environment for survival.

(Man is also, of course, different from all other animals. Once, intelligence, speech and culture were thought to be unique to man. Now we can see that his uniqueness lies in his ability to subject his surroundings to his own influence, thereby turning the tables on nature – the environment becomes a product of the species.)

The conclusion that man was in effect an ape was a bombshell to Victorian Britain and to the world at large. It was in direct contradiction to the Biblical doctrine that man was made in God's image. To Darwin's credit, he did not himself become involved. He knew exactly what he had done and why he had done it and he was, in a sense, almost apologetic. His spokesman became Thomas Huxley who, upon reading the *Origin of Species*, said he wondered at his own stupidity at not thinking of it himself.

The headlong battle into which the Creationists and the Evolutionists were now pitched was fought on many flanks but the issue finally came to focus in a meeting of the British Association for the advancement of Science in Oxford during 1860. The commanding figure representing the Church was Bishop Samuel Wilberforce; facing him was Huxley. Darwin significantly stayed at home.

Wilberforce, ever confident of a quick and decisive victory, left himself hopelessly exposed when in conclusion he turned to Huxley and inquired, 'And you, Sir, are you related to the ape on your grandfather's side or your grandmother's side?' Huxley thought for a moment and then (somewhat ironically considering the nature of the debate) murmured quietly, 'The Lord hath delivered him into mine hands.' Rising purposefully to his feet, he summarized the arguments for evolution and concluded: 'If I had the choice of an ancestor, whether it should be an ape, or one who having scholastic education should use his logic to mislead an untutored public, and should treat not with argument but with ridicule the facts and reasoning adduced in support of a grave and serious philosophical question, I would not hesitate for a moment to prefer the ape.'

Throughout the debate a lone Bible-clutching figure stood at the back of the hall. As Huxley demolished Wilberforce, the man lunged forward in an uncontrollable rush of emotion and made as if to hurl the book at Huxley, then thought better of it and fled the hall. It was Captain Fitzroy. The poor man must have felt himself largely responsible for the heresy that was now being unleashed upon Victorian England.

The debate on man's origins continues to rage, though among scientists rather than theologians. A mass of fossil remains have been found to justify Darwin. The evidence does not suggest – and never has, despite Huxley's and Wilberforce's slanging match – that we evolved from any ape that we know today. It suggests that both we and these apes had a common ancestor which separated into different groups, going their different evolutionary ways, something like 20 million years ago, during the Miocene Period.

Analysis of the protein molecules of blood among living primates suggests, on the basis that the more similar they are from different animals the more closely related those animals are, that this divergence may have taken place

Thomas Huxley – 'Darwin's Bulldog' (top) – crushed Bishop 'Soapy Sam' Wilberforce, the most ardent of the anti-Darwinians (above) at a famous debate on evolution in Oxford in 1860.

only about four million years ago. This claim, although it needs more thorough substantiation, does not actually dispute the fossil evidence for the *order* in which the apes and man may have evolved (gibbons–orang-utan–gorilla–chimpanzee–man). A significant result of these blood tests was that in the chimpanzee and man some 99 percent of proteins are identical, confirmation of the closeness of their evolutionary relationship. This similarity also indicates that man did evolve in Africa (because chimpanzees had never lived anywhere else), as Darwin had speculated more than 100 years ago.

The story that has emerged so far from the fossil record is still obscure, for the number of bones found are not enough to establish a firm sequence. But it is at least possible to present a tentative line to man.

About 30 million years ago, during the Oligocene, the first ape-like (but probably more monkey-like) creature emerged from the Old-World monkey stock. Its remains were found in about 1960 by Dr Elwyn Simons at the Fayum Depression in Egypt where once lush forests clothed the soil that is today on the bare, eastern edge of the Sahara Desert. The owner of the incomplete skull – for that was all the find entailed – was named *Aegyptopithecus zeuxis*, and it is regarded as the ancestor – or at least an ancestor – of all the living apes and man of today.

The next link in the chain – though it had actually been discovered some 12 years earlier by Mary and Louis Leakey on an island in Lake Victoria – was finally classified as *Dryopithecus africanus*. This fossil represents an ape some eight million years closer to us than the Fayum discovery and although it was still rather monkey-like in appearance, it was well advanced in the ape lineage. It was a creature that roamed far and wide in the forested world of the late Miocene and early Pliocene, colonizing a vast area including present-day Africa, the middle East, India, Pakistan and Europe.

Can we say in which of these countries this first real ape evolved? Intuition points strongly to Africa but the fossil record is still far too scanty for that continent to be marked down as a certainty. The only real certainty is that *Dryopithecus* (formerly called *Proconsul* after a chimpanzee named 'Consul' in London whose ancestor it was hoped it might prove to be) adapted to many regions of the earth and evolved several species. In this respect alone it showed great genetic flexibility.

The fossil record indicates that *Dryopithecus* was still in existence in Africa some 12 million years ago. So too was *Ramapithecus*, representing a new line in ape evolution. The most likely explanation is that *Dryopithecus* was to give rise to the gorilla and chimpanzee lineage in a direction that was dictated by the forests and that *Ramapithecus* was just a little bit different. For some reason – perhaps competition for food and space in forests already shrinking back towards the positions they occupy today – this latter ape found itself on the very edge of the savannah woodlands that were being exposed as the forests retreated. Life on the savannahs has always been much more of a problem for animals than has a forested existence. *Ramapithecus* apparently faced the challenge. This fact alone may have been the clue to the eventual emergence of man. The separation of the ape-trail to the family *Pongidae* and the human trail to the family *Hominidae* had taken place.

Ramapithecus was *essentially* an ape of the forests. Its fossil remains – no more than jaw and tooth fragments – have now been found in Africa, Europe and the Far East and their sites show that it probably lived on the edge of forest and more open spaces created by the action of large rivers. In the open areas, seeds of grasses would have replaced the fruit of the forest

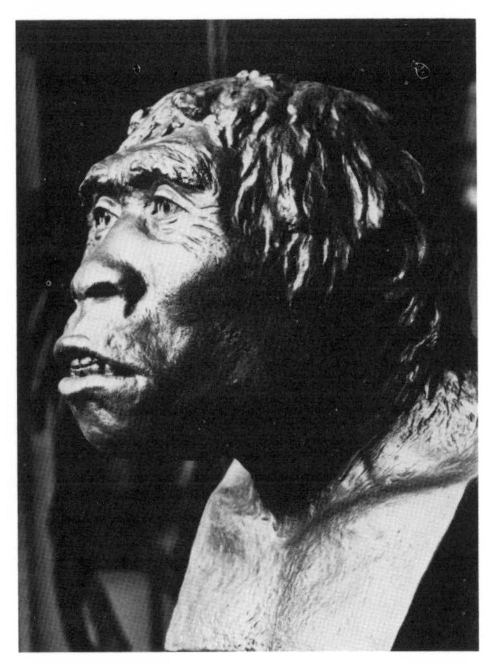

These early reconstructions are of what is now known as *Homo erectus*. When the first find was made by Eugene Dubois in 1891 (the model above was made by Dubois himself), this hominid ancestor was tabbed *Pithecanthropus erectus*, or simply Java man.

as the most abundant food source. The teeth of *Ramapithecus*, with their reduced canines in front and their enlarged, flat-faced molars behind, suggest such a change in diet. They are designed for grinding seeds rather than tearing at soft fruits.

It remains very speculative to suggest that ancestral man went through an exclusively seed-eating phase as he adapted to a new habitat. The dangers of loading so much responsibility on to single characteristics are exemplified by a creature known as *Oreopithecus* which was found in the coal deposits of Tuscany, Italy, in 1872. *Oreopithecus*, for whom some 50 good specimens now exist, shows a complete mixture of characteristics which encompass several early pongid and hominid features. It lived some ten million years ago, yet it is generally accepted as being no more than an evolutionary dead end which cannot be satisfactorily classified with either ape or man. It is important because it suggests that apparently hominid features were appearing, and being lost again, independently, in more than one group of Miocene primates. For this reason more *Ramapithecus* material is required before a conclusive picture of its life style and position on the tree of life can be established.

Whatever the evolutionary pathway, our ancestors did leave the forests and face the dangers of the open savannahs. Somehow our three-and-a-half-foot high ancestor had to protect himself from predators while enjoying his new-found food supply. It is possible that this necessity enforced further changes in ancestral man.

Some intriguing experiments by a Dutchman, Adriaan Kortlandt, give an insight into how early man may have adapted his behavior – and hence his physique – to increase his chances of survival in the new environment. Kortlandt compared the responses of forest and woodland (i.e. edge-of-the-forest) chimpanzees to the most feared predator of all (modern man excluded) – the leopard. He used a stuffed leopard with a mechanically operated head and tail to simulate life.

The forest chimpanzees reacted by tearing at lumps of wood and brandishing them around in a fearful display of defiance. They leapt into trees and shook branches while screaming and hurling broken-off tree limbs in the general direction of the leopard. In short they put on a tremendous display to distract and deter.

The woodland chimpanzees, however, went even further. They not only grabbed for dead branches but they also used them to club the leopard's body, with a force strong enough to inflict serious damage. This factor alone meant that they had to approach their enemy more closely than did their forest counterparts. But Kortlandt also noticed that they seemed to attack in more orderly groups as well. The final assault on the leopard consisted of five chimpanzees who encircled its body; their leader then grabbed at its tail and swung it with such strength that its head flew off to one side. Satisfied that they had won the battle, the chimpanzees settled down to a normal routine, although the head of the leopard was subjected to sporadic attacks for several hours to come.

It must be acknowledged that their response to the leopard may have been influenced by the fact that it *was* actually dead rather than alive. But the response suggests that in the savannahs, with their remote hiding places and their more widely-spaced food supplies which demand more travelling on the ground, the best policy of defense may be planned attack.

This policy in turn required better all-round vision, greater manipulative ability and more co-operation between individuals. These factors produced an increase in brain size, a more upright and two-legged stance and an increase in the use of the hands, which were used less for locomotion than ever before. And having evolved the means of protecting himself against hungry predators, early man would then have been in a position to use the self-same weapons and ploys as a means of attack. As is the case among so many hunting animals of today, one of the best ways of obtaining meat is to search for it in well-organized groups. Meat could be added to the list of foodstuffs.

It is conceivable that even at the *Ramapithecus* stage of development, the new and more open habitat enforced this kind of co-operation between individuals. They would all have benefited from such a formation, especially when repelling predators and hunting prey too large for any individual to tackle alone. If this was indeed the case then it possibly suggests that tool making – especially weapons which increase the powers of single hunters – was still at a very immature stage. There is certainly no evidence to indicate that tools were being fashioned by the ape-men of these early days.

This picture of life more than ten million years ago is supported by little hard evidence. It can be flushed out by working back from information on ape behavior available today, but until there is more evidence the behavior and appearance of our earliest human-like relatives will remain speculative.

The best era for the study of human evolution concerns only the last five million years or so. This is a big jump in time from *Ramapithecus*, but it is a well-documented period whose earliest fossils confirm that *Homo* was well on the way to becoming *sapiens*.

In 1924 Raymond Dart, Professor of Anatomy at the University of Witwatersrand, South Africa, discovered a small fossil head at Taung in Botswana. It was a large gorilla-sized braincase behind the face the size of a human child. He was convinced that it was something special in the ape-to-man lineage, a view largely rejected at the time, mainly because of the obvious immaturity of the specimen concerned. Dart, sure he had unearthed a new species, called it *Australopithecus africanus* ('Southern Ape from Africa'). In so doing he took the first steps towards establishing the actual stages between us and our earliest ancestors. His proposal was not popular: Asia was generally regarded as the homeland of early man, yet here was Dart proposing a completely different place of origin on the strength of one incomplete and immature specimen. The 'Taung baby' or 'Dart's child' as it came to be known was declared too much of an ape – and not a very special one at that – to be considered a *Homo* ancestor.

For Dart the cause might well have been lost but for the timely intervention of Robert Broom. Broom studied the Taung specimen some years later and pronounced himself heavily in favor of Dart's hypothesis. He was fascinated by the teeth, which lacked the large canines so typical of the apes, and further concluded from the angle at which the head must have joined the body that *Australopithecus* even walked upright on two legs. In 1936 the 70-year-old Broom arrived at sites near Johannesburg to search for more specimens. He found what he was looking for, with an unexpected bonus. He found hundreds of Australopithecine fragments, enough for him to announce that *two* types of *Australopithecus* ape-men were involved – Dart's small, slender *Australopithecus africanus* and his own solid *Australopithecus robustus*. He declared the finds to be about two million years old.

While the focus of attention was resting on the southern portion of Africa – although many people still thought it too isolated to be the center

This skull found by Robert Broom at Sterkfontein, South Africa, is that of *Australopithecus africanus*, one of the two species of hominid 'cousins' who co-existed in Africa some five million years ago.

This skull of a 5-or-6-year-old child – the Taung baby – was recognized as an *Australopithecus africanus* fossil by Raymond Dart in 1925, but it still remains one of the finest hominid relics.

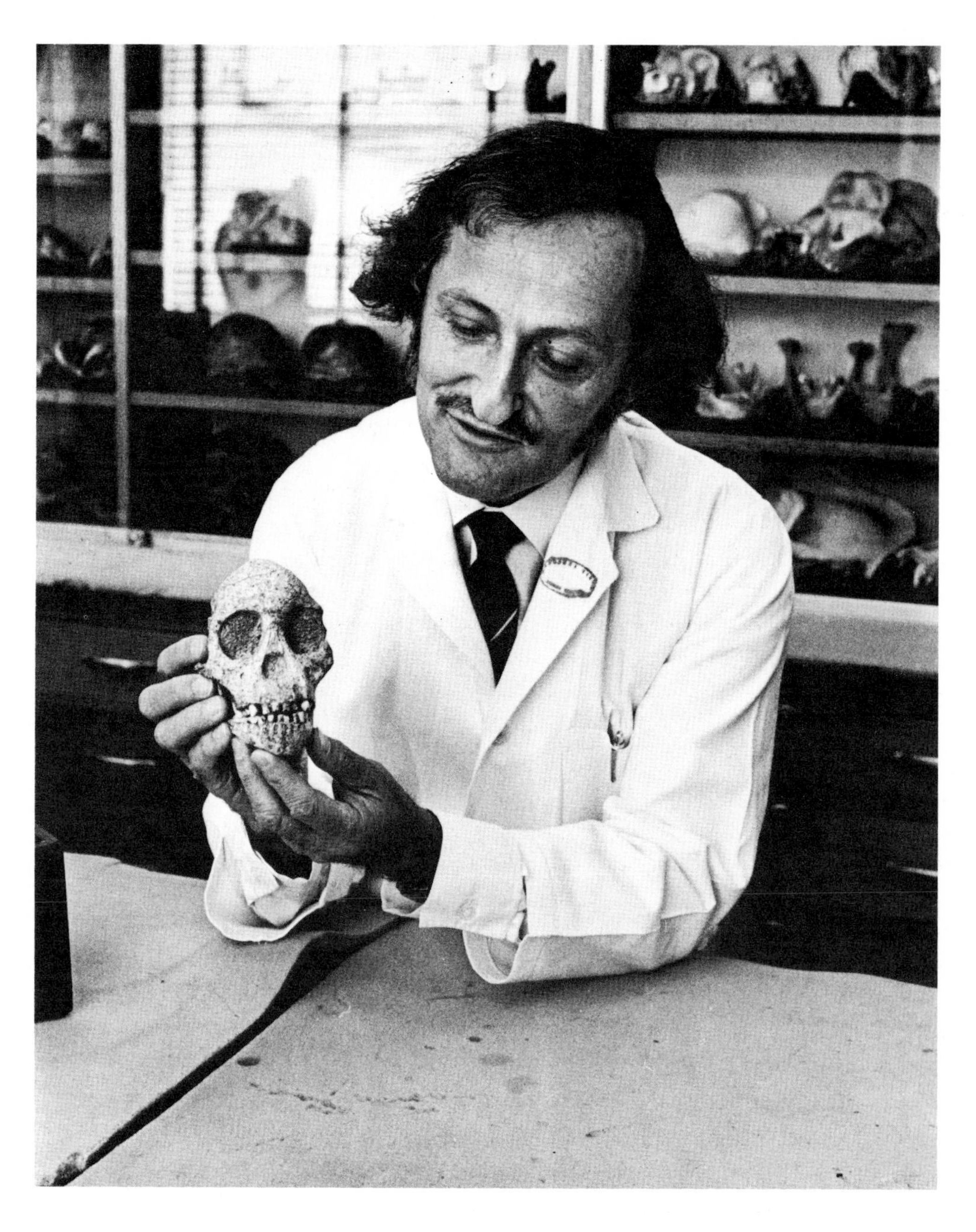

Dr Louis Leakey spent nearly 40 years searching for fossils in East Africa. The finds made by him – and his wife, Mary, and son, Richard – at Olduvai Gorge, Tanzania, have revolutionized our knowledge of our past.

of human evolution – Louis Leakey was following up a few clues in Tanzania to the north. The area of special interest to him was Olduvai Gorge where water erosion had exposed several layers of deposits accumulated over millions of years. The geological importance of this 25-mile scar on the eastern edge of the famous plains of Serengeti had been recognized as early as 1913 when a German, Hans Reck, produced from it an impressive series of animal fossils. Among his specimens were the remains, dug out from a low and old layer, of a human being. As it happened, it proved subsequently to be of quite recent origin – the body had been buried in the exposed earth by more advanced human types. Nevertheless it fired the imagination of Mary and Louis Leakey and in 1931 they began their most serious work in the gorge.

For 28 years they found little. It was not until 1959 that Mary's keen eye detected tiny fragments of skull protruding from the ground on the lowest and oldest level of the gorge. She quickly uncovered some teeth as well and knew immediately that she had found what they had been looking for all those years. It was a human of extreme age. More than 400 pieces were carefully assembled to form the nearly complete head of a male.

Louis Leakey saw differences between it and *Australopithecus robustus* and named it *Zinjanthropus boisei* ('Zinj' being an old name for East Africa). But the one and three quarter million year old specimen was soon recognized as an even more robust variation of *Australopithecus*. (The species name *boisei* was subsequently transferred to that genus.)

Since that day of discovery, Olduvai Gorge has also provided the first representative of the genus *Homo* – *Homo habilis*, identified partly on account of its smaller size but mainly because of its modern looking teeth.

The pattern of events was becoming clearer with every discovery. It appeared through the 1960s that late *Ramapithecus* had divided into different groups and that they had all lived at about the same time. In our quest for an orderly progression of evolutionary events, it is tempting to think of *Ramapithecus* giving rise to three Australopithecines – *A. africanus*, *A. robustus* and *A. boisei*. While the last two became extinct, *A. africanus*, the most lightly built of the three, gave rise to the more sophisticated *Homo habilis*.

But in 1972 Richard Leakey, the enthused and committed son of the pioneering family, was working at early human sites on the eastern shore of Lake Turkana (formerly Lake Rudolf) when Bernard Ngeneo discovered the remains of a skull of sensational importance. This famous '1470' skull – identified cautiously by only its museum catalogue number – proved not only to be about two and a half million years old but also that of an advanced *Homo* (probably *habilis*) type. The find indicated that *Homo (habilis)* was a contemporary of the early Australopithecines and that it may well have evolved separately from them rather than out of them. By this argument all the *Australopithecus* species would have to be considered evolutionary dead-ends and the southern Africa specimens, once thought to be the vital links between apes and man, were something apart from the main events. Perhaps the emerging *Homo* species had forced them out of East Africa – although they managed to live side by side for several hundred thousand years – and into southern isolation. But what was the ancestor of *habilis?* Was it *Ramapithecus* or some 'missing link' yet to be discovered?

Homo habilis gave rise to the next rung in the ladder to man, *Homo erectus*, perhaps as much as one and a half million years ago. By 750,000 years ago this latter man, in full control of its upright stance at last, was the sole survivor of this complex of human types. *Homo erectus* developed a sophisticated tool-making industry which included axes, choppers, chisels and various scraping implements. These could have only been the result of a superior intelligence which finally emerged in the face of competition and

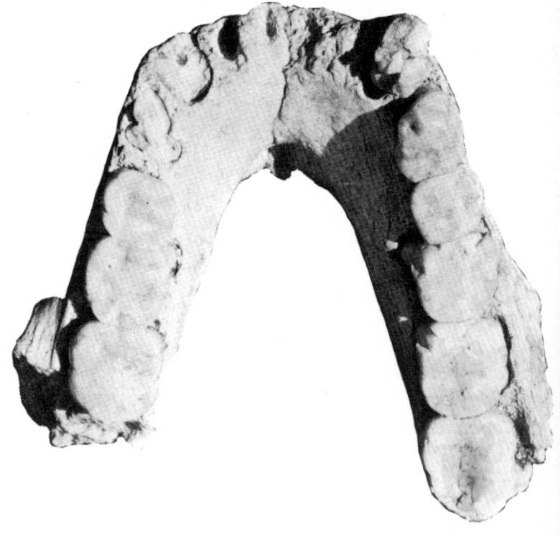

The jaw-bone of an Australopithecine found in Laetolil, Tanzania, dating back some 3.75 million years, is one of the tantalizing pieces of fossil evidence with which paleontologists attempt to unravel human pre-history. The evidence is still frustratingly fragmentary. All the fossil remains of man and his close relatives – a selection of which is shown at right – could be gathered on a desk top.

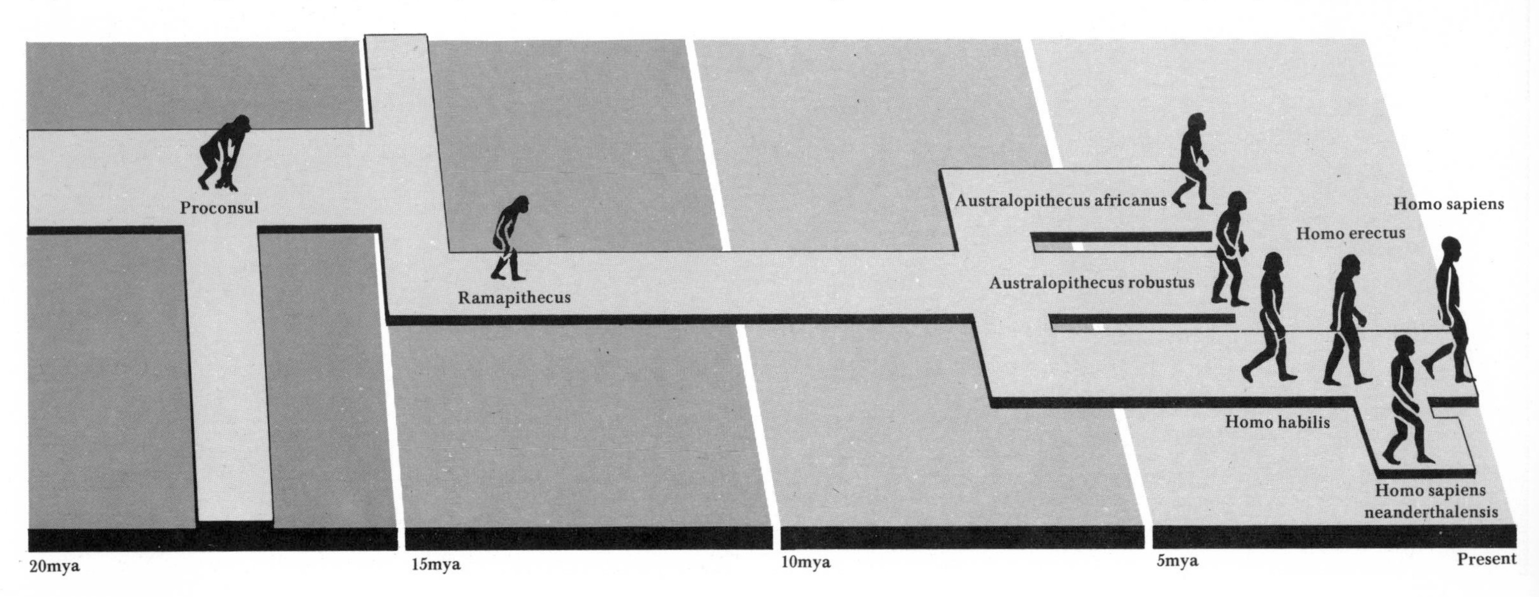

The latest of many patterns to be made of the fossil evidence, this diagram – a detail of the overall view on pages 14-15 – summarizes current thinking about hominid evolution. At one time, about five million years ago, it seems likely that man's direct ancestor was sharing his African homeland with three other hominids – two Australopithecines and Ramapithecus – all of whom died out. *A. robustus* is now considered to be the same as *A. boisei* (itself once known as *Zinjanthropus boisei*).

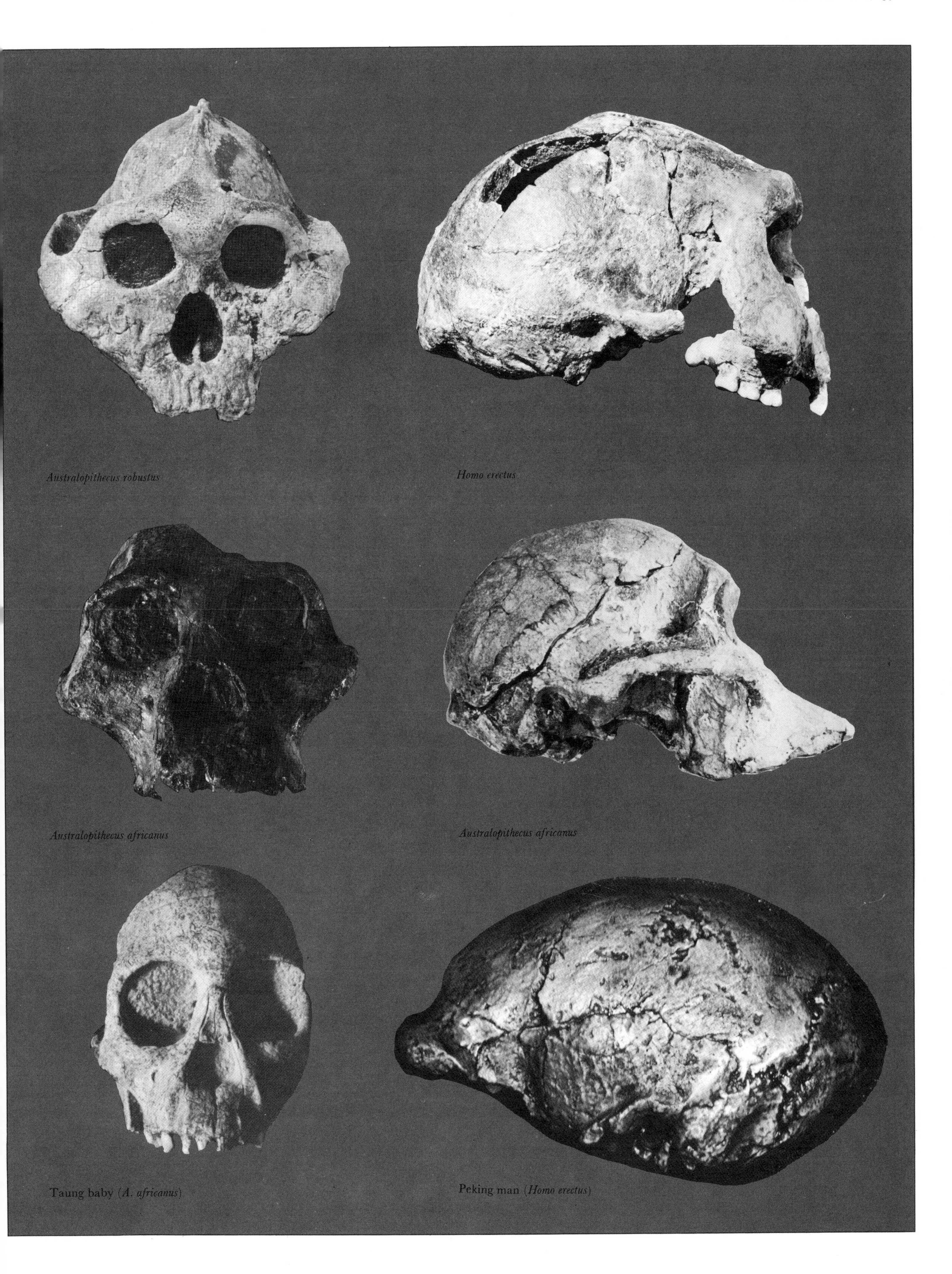

Australopithecus robustus

Homo erectus

Australopithecus africanus

Australopithecus africanus

Taung baby (*A. africanus*)

Peking man (*Homo erectus*)

Olduvai Gorge was carved out by a long-vanished river to expose ancient deposits originally laid down by a lake. An excavator points out the level at which an *A. robustus* fossil was found.

environmental consequences to drive all other closely related types into an early evolutionary grave.

If our *erectus* did arise in Africa, then it was at about this time that he began moving north and across the globe. His hands were as fully developed as our own, and he could transport food, water (perhaps in the form of succulent foodstuffs) and scant belongings. Although there is no hard evidence to suggest why man lost his hairy body covering, it is possible that it happened in response to an increased need to sweat once hunting animals on foot became an established way of life.

But as early man moved north toward areas covered in sheets of ice, he found need for protection against cold. Such conditions would normally act as sufficient barriers against animal dispersion but man, by now no ordinary animal, was able to preserve his body heat with skins from the animals he had killed for meat. And he gradually learned to put fire to good use.

Fire occurs naturally through the action of volcanoes, lightning and spontaneous combustion. In warmer climates man would not only have suffered by it but would also have profited from it. Natural bush-fires would have presented him with the cooked remains of perished animals. Burnt-off

A FALSE IMAGE OF MAN'S EXTINCT COUSIN

In the summer of 1856 workmen in the steep-sided Neander valley near Düsseldorf, West Germany, blasted open a small cave and came upon some ancient bones. They included a skull belonging to Neanderthal Man, a sub-species of *Homo sapiens* who lived between 30,000 and 100,000 years ago. (The skull was actually the second of its type – the first, found in Gibraltar in 1848, was largely ignored.)

The reaction to Neanderthal Man was one of revulsion. The owner of the skull was diagnosed as 'brutish,' 'of a savage race,' and a 'pathological idiot' – an attitude apparently supported when the severely contorted remains of an old arthritic individual were found in southern France. Neanderthalers became fixed in the popular minds as the archetypal human ancestor, with fearsome beetling brows, thrusting face, stooped, lumbering gait, and a stocky, muscular body – the very embodiment of stupidity and malevolence.

We can now be sure that the Neanderthalers led a complex, thoughtful and sensitive existence. For instance, they buried their dead, as we do, with sprays of flowers. These people – not our direct ancestors – were specialists in coping with the harsh conditions of Ice Age Europe and succumbed only because they were too specialized to adapt to the warmer times in which *Homo sapiens sapiens* emerged.

These museum reconstructions and the two overleaf exemplify the now outdated view of Neanderthalers as brutish ape-men.

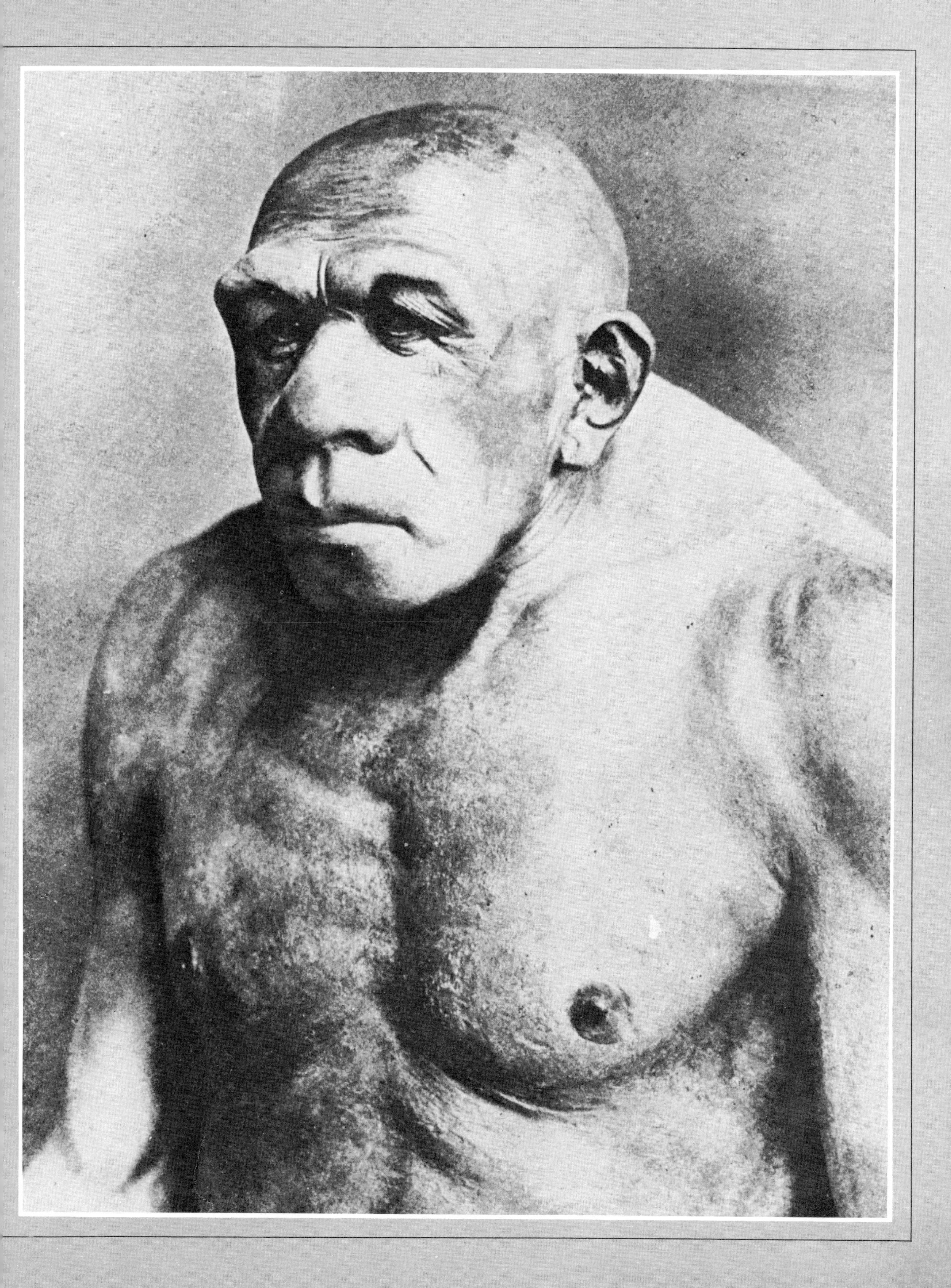

areas would have been replaced by succulent new growth ready for harvesting and eating.

The first evidence of man using fire in a more social sense does not come from Africa but from Europe, from Escale near Marseille in France where a hearth-containing cave was exposed by dynamite operations in 1960. The discovery opened up new interpretations of early life. The cave had obviously become the focal point of existence as a protection against the cold. And the cave with a fire in it was even better, for it provided immediate warmth, kept wild animals at bay – especially those intent on returning to the home from which they had been evicted – and, critically, it molded social life by providing a protected area that brought humans together.

The increased brainpower which led to these dramatic improvements in daily life entered a new phase as the benefits of verbal communication established themselves. Fossil records cannot testify to the evolution of speech but it can be speculated that man's increasing awareness of his surroundings was a significant criterion in this respect. He could fashion implements for different uses. The greater intelligence, the growing curiosity and the distinct advantages arising for those people whose sounds could be modified in a more meaningful way, led to the gradual evolution of our languages. It might even be reasonable to assume that speech arose in Europe where the conditions of survival and the benefits of improved life were so much more sharply felt in the face of adverse conditions than elsewhere. It must have been a very slow process, taking perhaps half a million years and culminating no more than about 50 thousand years ago.

By this comparatively recent date in Earth's history, man had made his impact on the world. One variety of our species, Neanderthal Man, in this case a subspecies, *Homo sapiens neanderthalensis* was coming to the end of a successful European existence in which he had adapted to ice-age conditions. The Neanderthals may even have over-adapted to such an extent that when the first phase of the last great ice-age came to an end, they could not survive under the improved conditions.

But living at the same time as the Neanderthals was another variety of human – also derived from a *Homo erectus* ancestry – which survived the change in climate. This was Cro-Magnon Man, physically identical to us. It has often been suggested that this must have been a bloody period in the history of man, with Cro-Magnon systematically annihilating the Neanderthals until they went extinct. This is probably not true, although competition between the two, where they overlapped, may well have existed. In some areas the two closely related types may even have interbred, giving rise to a Cro-Magnon-based human population containing Neanderthal genes within it.

Whatever happened, Cro-Magnon Man emerged as the dominant strain. It is his paintings which grace the famous caves of Lascaux in France and of Altimira in Spain and it is his implements which, for the first time, incorporate artistic design above basic use.

Cro-Magnon Man flourished across the globe, forcing his way into remote areas like Australia and America as recently as 10,000 years ago. His small family units of the past became large tribal concerns, and settlements replaced the more nomadic way of life. If there were any Neanderthals surviving anywhere, they must have been pushed by this onslaught into the remotest corners of the earth where they would almost certainly have perished.

By 10,000 years ago *Homo sapiens sapiens* had invented agriculture. With

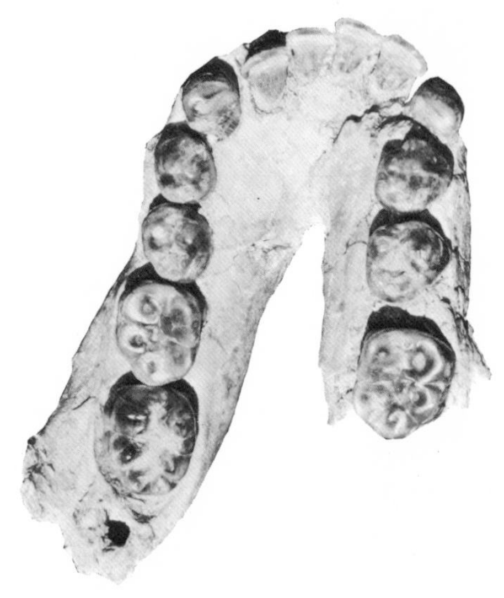

A 14-inch *Homo habilis* thigh bone (top) reveals that man's earliest direct ancestor walked fully upright – the bone is almost identical to modern man's. A *habilis* jaw, with its heavy molars (above) shows that the creatures were omnivorous.

Richard Leakey at the Koobi Fora site, Kenya, shows the painstaking work necessary to reveal the slithers of bone that, when combined, form fossils like the skull below.

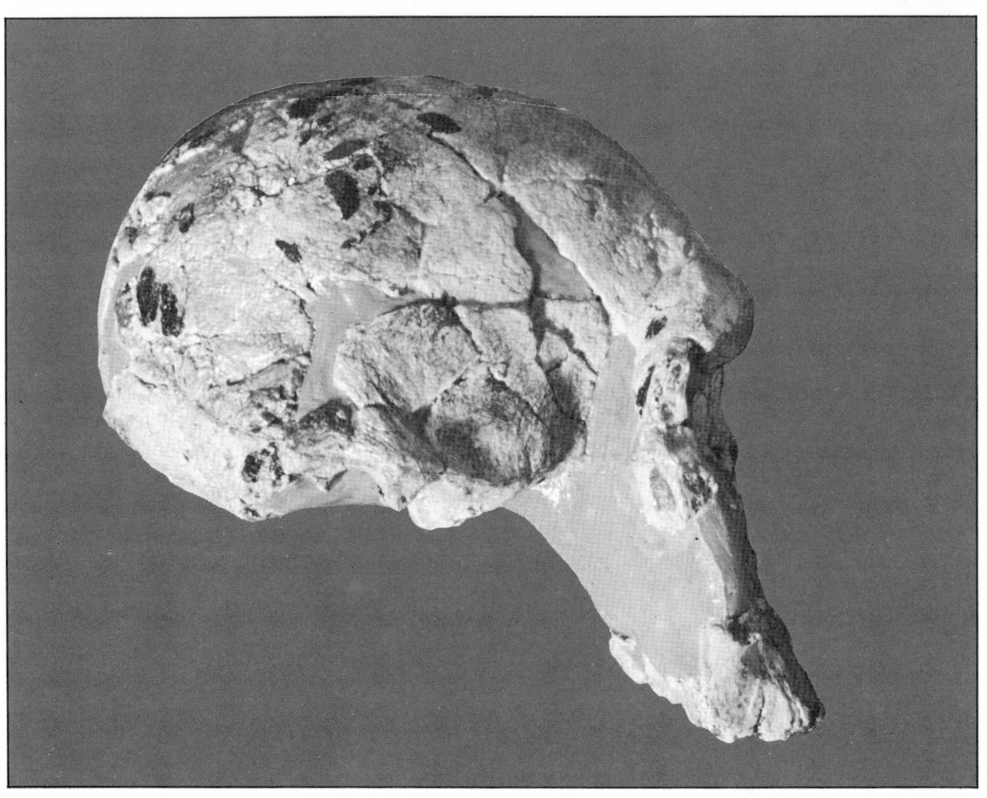

This *Homo habilis* skull – known by its catalogue number, 1470, in the National Museums of Kenya – is two to three million years old, proof that early *Homo* and the Australopithecines were contemporaries.

Charles Dawson (also inset, top) and Smith Woodward work at the Piltdown site.

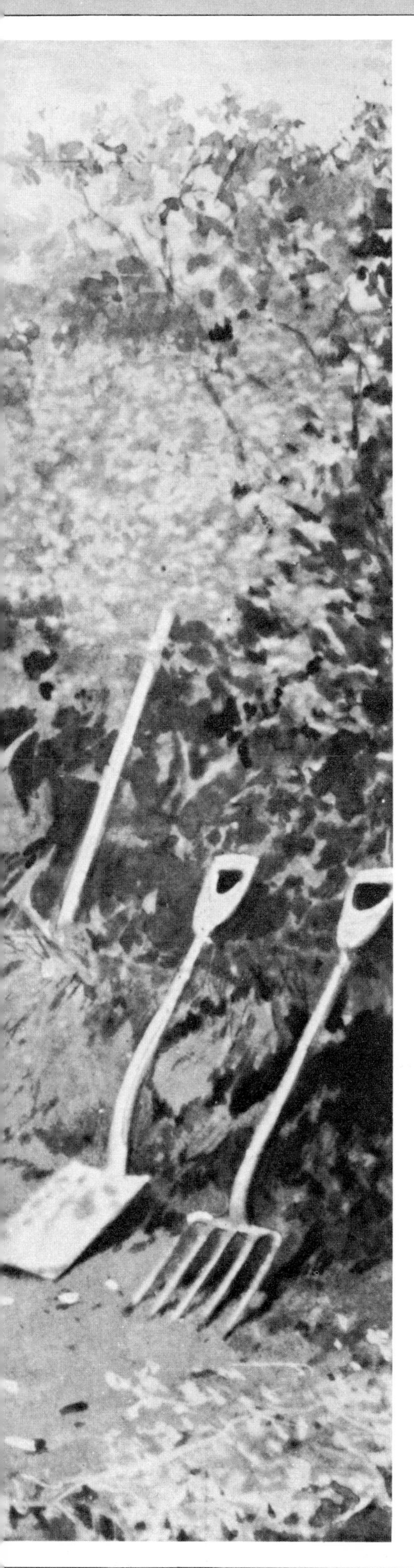

From Piltdown, the Man Who Never Was

The Piltdown Man was ranked as one of the most successful scientific hoaxes ever. In the early years of this century, English archeology was at a low ebb. Major discoveries – of Java Man and Neanderthal Man – had been made by Dutch and German scientists. It was widely believed that *Homo sapiens* must have evolved in Europe or Asia.

When in 1912 Charles Dawson and Smith Woodward unearthed fragments of an apparently semi-human skull in a Sussex gravelpit near Piltdown, the supposed creature was eagerly welcomed into the family of man as the 'missing link.' Persuasive ape-man reconstructions – like those above and below – flooded the popular press.

The results were scientifically disastrous. Raymond Dart's identification in 1925 of the important Australopithecine 'Taung baby' was largely ignored, for few accepted that Africa could have been the cradle of mankind.

Only 40 years later was Piltdown Man shown to have been a carefully doctored combination of a modern cranium and an ape's jaw. The forger is still unknown.

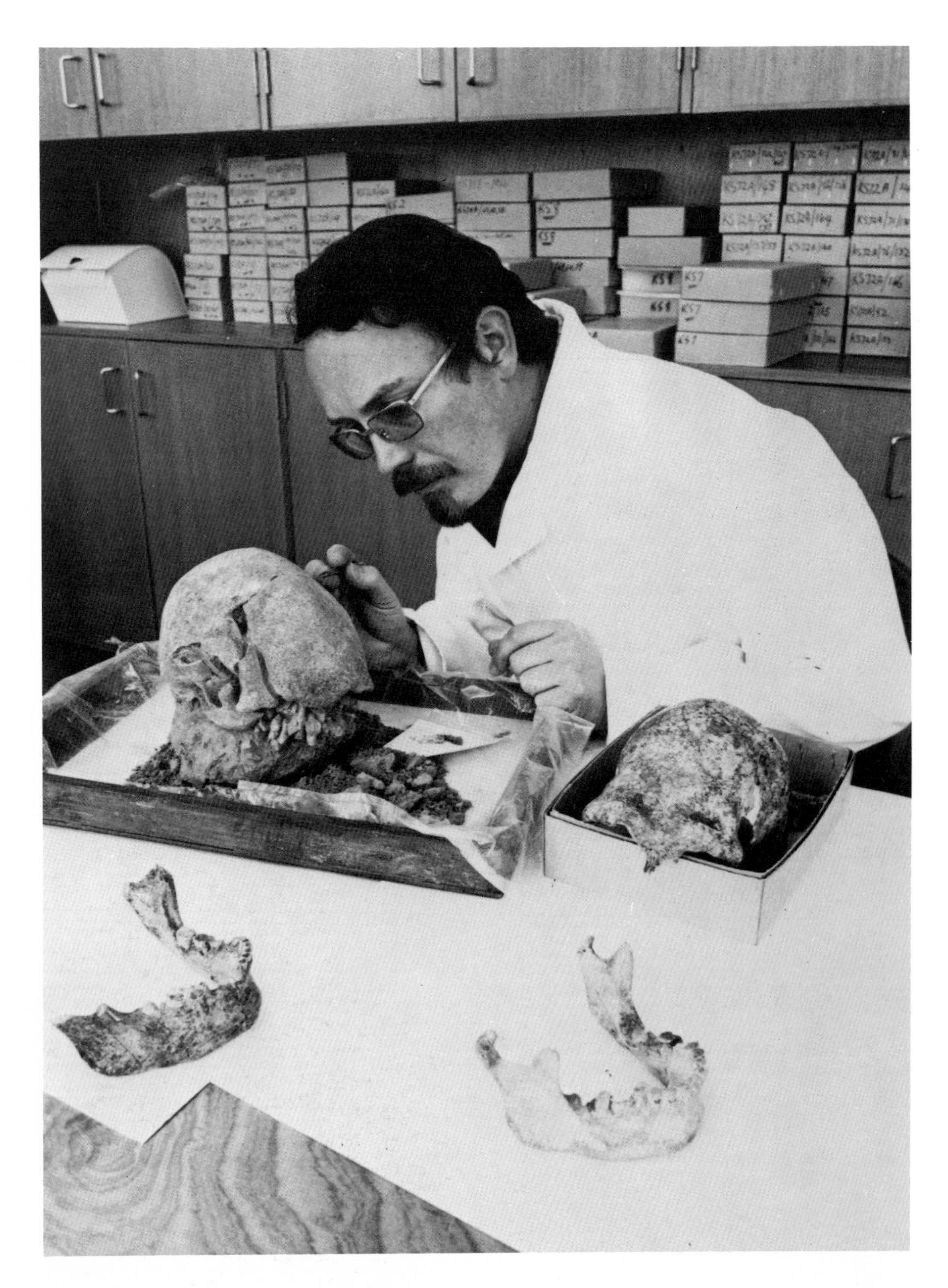

This find from Australia shows that many details of man's spread across the face of the Earth have yet to be explained. The bones were found in Kow Swamp in Victoria, and are some 10,000 years old; but they apparently retain some features of *Homo erectus* – long after the evolution of the successor species, *Homo sapiens sapiens*. It is possible that Australia, as a geographical backwater, supported both species at the same time.

it came the need to protect property. Rituals and ceremonies associated with birth, puberty, marriage and death developed. Cultures and societies diversified across the globe as men protected themselves more and more from the biological world. Agriculture gave way to industry. Now man can dominate the forces that once controlled his evolution.

We now stand at a momentous point in our evolution: within a few decades we shall have established permanent bases in space. If so, it will be a logical continuation of our increasing independence – from the protective parent, from the vagaries of natural food supplies and the presence of predators, from the constraints of climate and now, finally, from Earth itself.

In some ways the quicker this occurs the better. Only then will we be free to ensure the survival of those areas of the Earth that contain vital clues to our own origins – the homelands of the other great apes with whom we share a common ancestry.

THE GREATEST APE EMERGES

Between 50,000 and 10,000 years ago, *Homo sapiens sapiens*, the particular subspecies to which we belong, spread inexorably over the globe. In Africa, Europe and Asia, man continued the way of life established by his *Homo erectus* predecessors, combining to hunt other creatures for his own betterment – for food, for clothing, for the artefacts that formed the basis of civilization. In the icy north, mammoths supplied meat, skins and ivory (below). Some 26,000 years ago, man moved across the Bering Straits into the Americas.

Eventually, as warmer conditions brought an end to the last Ice Age some 10,000 years ago, a number of tribes independently discovered the advantages of settled life. Farming, agriculture, the establishment of permanent dwellings – and civilization as we know it – began, and man's adolescence ended.

Early men throw bolas in a baboon-hunt.

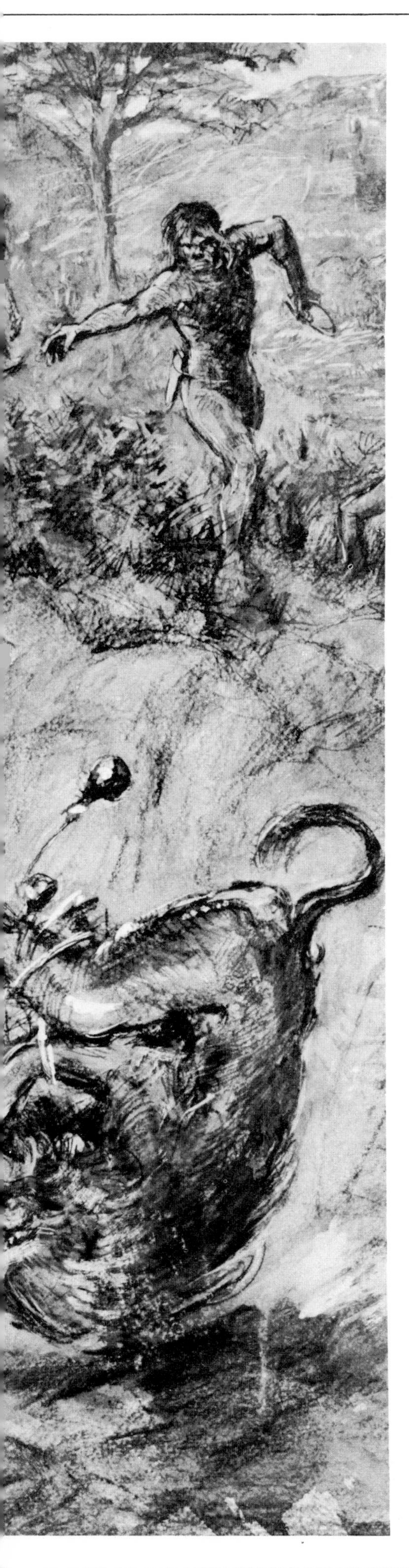

On the African savannah, a tribe butchers a wart-hog.

6/PROTECTING THE THREATENED APE

The success of one of the apes – *Homo sapiens* – is inexorably undermining the existence of the others. Poaching for zoos, hunting and the destruction of habitats all combine to threaten the future of our closest living relatives. In some areas, further intervention has had limited success – like the rehabilitation schemes in Borneo and Sumatra in which orangs may be drugged (left) prior to receiving an education designed to allow them to return to the jungle. But in the long term only the preservation of the habitats, to the advantage of both animal and human populations, will ensure the apes' survival.

The extinction of species is a natural phenomenon in the world. So, too, is the formation of new ones. As world climates change, as land masses break up and drift apart and as vegetation structures alter their appearance, so the animal life that has evolved with them undergoes changes as well. Today's one million or so species represent a mere two percent of the species that have ever existed on Earth.

One of these – man – is unique. He can control his own environment, and does so to such an extent that he is at present facing a crisis of his own making. Numbers grow, demand for food grows, resources – including available land – dwindle. We may find a way out of this crisis with 'zero population growth' and new sources of power or even space migration. But whatever ways we find to survive, pressure on other species, in particular our own evolutionary relative, the other apes, will diminish only as the result of conscious, planned conservation.

Conservation is often regarded as a policy to protect individual species. In fact the concept is far more wide-ranging, involving the protection of the habitats within which species evolve and live. In the widest sense, we have to conserve ourselves – protect our own habitat. That is what the environmental crisis is all about. But as a consequence we have to protect smaller scale areas as well – in particular, the rainforests, which supply much of the Earth's oxygen.

When placed into this sort of context, the plight of the great apes today becomes rather acute. Their numbers may be dwindling because of com-

Deforestation – like this in Borneo – is the greatest threat to the continued existence of orangs. Deprived of their habitat, they die from lack of food, either directly or as victims of unnatural competition as populations are crammed into ever smaller areas.

mercial exploitation – which expresses itself in a multitude of ways from killing for trophies and trapping for live export to interference from over-zealous tourists –but of what use is it to campaign for the orangs when the forests in which they live are being felled out of existence? Already local climates have been altered as a result of such large-scale deforestation. The great apes, locked as they are to a forest existence, represent something of paramount importance to us all. If we can preserve them in their intact habitats, we increase our own chances for survival as well.

The preservation of global habitats is an international issue, particularly involving the developing countries which almost by definition have more areas of wilderness in them than developed countries. Wilderness areas represent resources. Are such resources to be exploited? Or left untouched – to the economic detriment of the country concerned?

Such problems demand a worldwide solution over many decades. In the meantime conservationists can fight to maintain populations of wild animals at safe levels – levels at which they can establish normal social and breeding relationships – so that when the day of full-scale habitat protection arrives, they are still living in sufficient numbers to enjoy an active and undisturbed future.

There are still, fortunately, people who are prepared to spend their days working towards a better world for wildlife – people who, often at great personal risks to themselves, seem willing to take on the almost impossible. Above all they must fight other people who are motivated as much by a desire to help their own countrymen as by greed and ignorance.

The Saving of the Orang-Utans

It was on Christmas Day, 1956, that Barbara Harrisson received her first orang. Its mother, typically, had been shot and the abject one-year-old removed to a native house. There it had been found by a Bornean forest ranger who had confiscated it and passed it on to Tom Harrisson, Curator of the Sarawak Museum. He had brought it home to his wife as a Christmas present. The helpless young male – named Bob – would surely have died if not properly cared for. The love and attention that was lavished upon him brought him safely through the danger period of vulnerability to human disease and malnutrition from eating the wrong food. His survival was guaranteed, albeit in a situation far removed from his native jungle haunt.

Bob was shortly joined by a young female, Eve, then other orangs who had suffered a similar fate. With this influx of animal life into their home, the Harrissons found themselves committed to a policy of raising orphaned orangs in captivity. Even in the earliest stages of their venture they could foresee the problems ahead. In particular, what was to be done with the youngsters when they grew too big and strong to be housed with the family? Orangs develop slowly, learning the arts of jungle survival from an experienced mother. These could not be learned in a human household. Return to the jungle would mean death.

By 1958 Bob was three years old and too well adapted to home life to stand any chance of survival in the forest. He was flown out to San Diego Zoo, California, where he was a great success with the American public. The female Eve, as dependent as Bob upon humans, was sent off to Berlin Zoo where she lived contentedly in the new ape house. Barbara Harrisson knew that these decisions could not set a precedent. The mothers were dead and now their offspring were living thousands of miles away in captivity. Even if they did produce young, the chances of them contributing to the wild population were virtually zero. What was really needed was a program by which the orphaned youngsters could be returned to the wild. It would mean training them for an orang life rather than for a human life from the very first day they were found.

When the Harrissons received their next consignment of orangs – word had soon spread that they were the people now responsible for them – the youngsters were not treated so softly. They were forced to climb trees to find fruit and were not doted upon as their predecessors had been, in the hopes that they would develop their natural tendencies sufficiently to be returned to the forest in the future.

The three new males, Bill, Frank and Nigel, were willing and able to treat the trees of the Harrisson's town garden as a natural home. They learned to pick fruit and to build nests. Nigel, especially, would spend the night outside in one of his branch shelters. The main problem was that in bad weather they would come down to the shelter of their cages where they also found additional supplies of food and drink. They may well have become half-wild but it would also have been impossible for them to relinquish their dependency upon humans. As the limited supply of trees in the Harrisson's garden were stripped of their fruit, so the dependency

Teaching an Orang Jungle Ways

'The return to Kuching after three weeks' exploration in the Sebuyau area was celebrated by way of a hot bath and luxurious relaxation in comfortable chairs. We had learned and seen much in the jungle and had followed up every piece of information. Yet we had made no essentially *new* observation. Unable to follow any of the animals for long, we had had no glimpse at mating, at mothers, at social life, organised groups. The dream of seeing and understanding some of the fundamental attitudes of one of our closest relations in the evolutionary tree had become something of a fantasy. The Orangs had been stronger than we in their own world of trees.

But the trip toughened my approach. I made up my mind to try and be a true ape-mother from now on. The children would have to spend much time in the trees instead of on the ground. Perhaps they could learn to build themselves nests. Anyway, they should live as free of me as could practically be.

Eve was a big girl now; two and a half and as jealous as ever. Bill had grown quite fat. He had already developed a pouch under his chin and Bidai reported that he was very greedy. Frank was much more slender and lively. All three greeted me with mild enthusiasm.

A group of fruit trees stood conveniently apart on a slope in the back of our garden.

"From now on," I said to Bidai, "put them *into* these trees every afternoon. Don't bring them down, even if they climb high. Let them do as they please."

He did not like the idea and protested: what were we to do if they tried to get away? I told him not to worry. We must see for ourselves, I said.

Eve was fearful at first. Every time Bidai put her in a tree she came down again to be cuddled. I teased him:

"You have spoilt her too much! Look; she behaves like a tiny baby, she cannot be without you!"

He was indignant. Had he not looked after her well always? Given her food, leaves to play with every day? Cleaned her cage, bathed her, never, ever smacked her?

"She must *learn* to grow up, Bidai, live in her own world, get away from babyhood. Don't allow her to cling to you, teach her to stay in the trees! Sit up there *with* her. Perhaps she will get used to it then!"

So he climbed half-way up the tree, and she went too, first clinging to him and then finding her own way.

It was easier with the babies. Frank behaved as if he had never been away from trees. He climbed swiftly, selecting slender branches and creepers where his small fists and feet could grip and went right to the top. There was no question of fright or even hesitation: he was clearly enjoying himself. Bill followed, but did not go high. Both fed on fruit and leaves and then played, swinging and chasing each other.

When Bidai came down from his perch later, both Eve and Bill followed to romp on the ground near him before he took them to their cages at dusk for their supper of milk, rice and fruit. He asked me what to do about Frank, who was least attached to him and independent by nature.

"He does not want to come down. Shall I go after him?"

"Let me try first," I said. "Perhaps he will make up his mind if I show him his supper."

It was nearly dark and I could see him sitting high up in the tree, a small black shadow. I took his feeding bowl and called:

"Frank . . . Frank! Come down for your supper!"

I sat where he could see me and called again. He squalked once and started moving down. Half-way down he stopped and sat watching me.

"Frank . . . be a good boy . . . come down to your mama!" I lifted his bowl.

Another squeak and he came, straight into my arms, to be cuddled and have his supper. He ate this greedily, right under the tree. I petted him and took him to his cage to bed. Bidai was impressed.

"I think he would not have come for me," he said. "He likes you better."

He was perhaps right. Frank, though I did not spend much time looking after him and seldom carried him about, seemed to have formed a spontaneous attachment to me. It became a ritual every evening for me to call him down. There was no need to take his feeding bowl. As soon as he saw me he started to come, straight down to the branches immediately over my head. When he was two or three feet above me he let himself fall into my arms.'

Barbara Harrison, *Orang-Utan*

increased. By March 1959 Bill, the least willing to give up a life of domestic bliss, was sent off to Antwerp Zoo in Belgium. By the following year both Nigel and Frank had demonstrated that they would never be able to live successfully on their own in the jungle and they, too, were sent off to a zoo life in Europe.

The Harrissons had had a moderate success. They had kept alive orangs which would have undoubtedly died, they had helped to furnish zoos the world over with breeding stocks – and there is nothing like the presence of an endangered species in a zoo to arouse public feeling for it. They had also shown that nothing less than a full-scale and independently financed rehabilitation scheme would be required to rear young orangs in such a way that they might eventually be returned to the wild. They could hardly undertake such a task themselves, but it was not long before such schemes were being put into operation.

In the early 1960s the North Borneo (Sabah) Game Department established a rehabilitation center at Sepilok between the Labuk and Kinabatangan Rivers. Its aim is straightforward enough – to return to the wild the young orphans who have been found in the forest or confiscated from people keeping them illegally as pets. The problems are immense because too many of the orangs have become overdependent upon humans before entering the center.

Once there, they must learn the ways of the wild from old and more experienced orangs in the naturally forested reserve. Given the long maturation period of these apes – anything up to ten years – it is obviously a long-term project. It relies also upon government legislation to maintain the forests in a healthy, orang-supporting condition. The first success of the Sepilok reserve occurred in 1967 when Joan, a rehabilitated female, went off into the forest and returned pregnant after meeting up with a wild male. She gave birth to a female, Joanne, who remained in the vicinity of the center until she was three years old. She then left of her own accord to fend for herself away from human influence.

On the south coast of Borneo at Tanjung Puting – in Kalimantan, that part of the island ruled by Indonesia – a similar rehabilitation scheme is under way. It began as a research center for the study of wild orangs and as such was occupied in 1971 by the young scientist Biruté Galdikas-Brindamour and her photographer-husband Rod Brindamour. They spent several years monitoring the behavior of Southeast Asia's vanishing ape and not only did they add valuable information to the pioneering work of John MacKinnon but they were also successful in reintroducing orphaned orangs to the wild.

In response to a request from the Nature Protection and Wildlife Management branch of the Indonesian Forestry Service to establish their research camp as a rehabilitation center, they incorporated several homeless orangs into their daily lives. The first, a male called Subarno, had learned much of the artistry of jungle survival before he had been captured as a three-year-old by poachers. With such a natural advantage conferred upon him, he was able to return to the forest after only a few months of camp life. But the similarly-aged male, Sugito, had been reared in captivity before being welcomed into camp and he clung to his new foster mother for more than a year, rarely attempting to move far away and even more rarely showing any signs of jungle self-sufficiency. He was joined by seven-year-old Cempaka, a female who had spent all her life being treated as a human baby.

As the years passed the orangs grew stronger and bigger. By 1974 there were four of them who had all begun camp life as human-dependent, badly nourished and underweight orphans. The time had come when their continued association with humans was no longer working to their advantage. They were ready for rehabilitation and the domestic ties had to be cut.

A feeding station was built for them some distance from camp and an ape-proof house was erected for the humans. The orangs were then left to their own devices. They destroyed the old hut where they used to sleep and, having done this, were forced to build their own nests in the forest trees. Gradually they moved further afield, obviously finding more of their own food, and their visits to camp became less frequent.

Within a few months several weeks would elapse between their visits to their former home.

For Biruté and Rod Brindamour the exercise had been more successful than they could ever have dared hope. Four or five new orangs recruited into a local breeding population could have an enormous effect on the long term numbers of individuals in the area. While these numbers are being patiently built up by only a handful of dedicated people, it is vital that the destruction of forests for commercial reasons be halted, if the efforts of the conservationists are not to be in vain.

Gorillas: Victims of the Trophy Hunters

Of the three subspecies of gorilla it is the mountain gorilla which has given so much cause for concern to conservationists throughout the world. As early as 1933 the Central African population was given complete protection

Woodcutters hack at the Sumatran rainforest (left) cutting swathes in jungle (right). To ban such activity – even if it were possible – would not be a complete answer, for the future of the orangs must be seen within the context of the need by individuals and the nation to harvest the forest as a resource.

under the London Convention on the assumption that its numbers were measurable in no more than hundreds. Fortunately the assumption was erroneous. George Schaller concluded that not only does it occupy a range far greater than was previously supposed but that it also occurred in numbers measurable in thousands. Today, however, hundreds may be a more accurate assessment.

The effects of man are such that the lush gorilla-supporting habitats are being quickly eroded by pastoralists and agriculturalists alike. Even the officially safe Virunga Volcanoes, housed within the Albert National Park, are under perpetual seige as the park boundaries are whittled away by the illegal destruction of the forests.

One other major problem is posed by the gorillas' habitat themselves. Gorillas survive well in primary forests at high altitudes away from human settlements, but they move willingly into areas of deforestation where a growth of secondary forests offers them a richer supply of succulent vegetation. Thus they come down to lower altitudes where they are effectively competing with humans. They become easier prey, both to those natives who still hunt them for food and to those who hunt them to sell as trophies to tourists.

The gorilla habitats border three countries – Zaire, Rwanda and Uganda. Both Zaire and Uganda have shown some interest in conservation, partly because of the tourist income such a policy can bring. Rwanda can barely afford its national park, and it is here that the gorillas are most threatened.

In response to this threat, Dian Fossey, an American scientist, has spent the last decade fighting to protect groups of gorillas against poachers, pastoralists and overzealous tourists. She is head of the Karisoke Research Station situated some 9840 feet above sea-level on Mt Visoke in Rwanda's Parc National des Volcans, part of the Virunga Volcano complex.

Here there is no adequate park-patrolling system, and wood-cutting, herd-grazing and poaching combine to make the task an almost impossible one for Dian Fossey and her small team. But at least she knows that her very presence in the area acts as something of a deterrent to the people who would willingly destroy her research area and the gorilla's livelihood. Her greatest enemies, only a few of whom have been caught and imprisoned,

A Program for Gorilla Protection

'As a result of our work, I believe that the following recommendations are essential to the preservation of the mountain gorilla:

1. The gorilla should remain on the list of completely protected animals. This means that every effort should be made to prevent the mass killing of gorillas for food by natives; that modern firearms should not be made available to them to facilitate such killing; and that local governments should place and enforce a strict limit on the number of gorillas captured by zoo collectors, medical institutions, museum collectors, and others. The killing of a female to obtain the infant should be outlawed.

2. The habitats in which gorillas live should be preserved, especially those in areas of rich volcanic soils where permanent cultivation is possible. A special effort should be made to prevent incursions by agriculturalists and pastoralists into forests which already are reserves and contain gorillas – the Virunga Volcanoes, the Kayonza Forest, the Mt Tshiaberimu massif, the Mt Kahuzi forest reserve.

The creation of large nature reserves in lowland rain forest at present uninhabited by man would aid in the perpetuation of the gorilla and other wildlife in the event that improvement in agricultural methods and a greatly expanding population disrupt the pattern of shifting cultivation. Such reserves would not need to be protected from all human disturbances, some forest cutting being actually desirable, for gorillas find optimum foraging conditions in secondary rather than primary forest.'

George Schaller, *The Mountain Gorilla*

are the poachers, who slaughter individual animals for purely commercial reasons. Poachers killed the last leopards and elephants in this region at least ten years ago. They have recently turned their attentions to the gorillas. The fate of two of Rwanda's patriarchal gorillas bears testimony to the poachers' ruthless and barbaric behavior.

Toward the end of 1977 one of Dian Fossey's best known gorilla groups contained a splendid young male whom she knew as Digit. He was twelve years old and had been known to her since he was two. She had witnessed his gradual rise through the ranks and had even grown to be trusted by him to such an extent that they could touch each other without showing any signs of fear. The group that Digit lived with was headed by a grand old silver-backed male called Uncle Bert. Together with their companions the two males were world famous, the stars of television programs and tourist posters.

Now they are both dead, murdered by poachers keen to supply an underground trade with prized heads to go on walls and hands to be used as ashtrays. Digit died on 31 December 1977 as he fought a desperate rear-guard action to cover the frantic retreat of the rest of the group. Uncle Bert died for the same reason on 24 July 1978, his speared and headless body being found by David Watts who followed a characteristic 'fear-trail' of smashed vegetation and dung for almost three miles through the forested slopes of Mt Visoke.

It is said locally that the people most likely to be responsible for these senseless killings are from the Batwa tribe, but what actually happens to the trophies is not known. Another possible motive for the killings is revenge against Dian Fossey for the imprisonment of several of Rwanda's most renowned poachers. One, Munyarukiko, Digit's murderer, committed that crime shortly after his release from prison.

The gorillas are threatened by such activities not simply as individuals, but as groups. Gorillas cannot function properly as a group without a strong social structure. When the big males – the best trophies – are removed, the whole group, now leaderless, suffers as well, with a marked reduction of breeding success.

Chimpanzees: A Rosy Present, An Uncertain Future

The chimpanzee enjoys the widest distribution of all the great apes. It is more adaptable across its range than are its larger relatives, and as such it does not usually warrant the concern allocated to them. But it will very soon give us cause to worry about its future, for its continued existence depends largely upon the fate of Africa's rainforests. Where agriculturalists convert non-forested lands of Africa into corn and maize monocultures, chimpanzees will be moved out of the way; where farmers grow acres of exotic fruits for export, chimpanzees will be persecuted for stealing valuable produce.

The pygmy chimpanzee, which may still prove to be our closest living relative, could disappear completely because it is locked into a forest existence. But although the common chimpanzee occupies a wider and more varied range it is not necessarily that much safer. Indeed so many of them are captured for medical research, export abroad and to be eaten locally that much more concern for their safety should be shown. The common procedure of killing females and removing their young operates on such a wide scale that Jane Goodall once calculated that for every individual *safely* exported at least six others die. Such wanton slaughter on even a mildly endangered species must soon have dramatic effects.

Game wardens display young orangs.

Capturing a young orang.

Among the several schemes to rehabilitate orangs is the one at Bohorok, Sumatra, backed by the World Wildlife Fund and the Frankfurt Zoo. Over the last few years, scores of apes have been taken in when young, cared for and finally weaned away from their dependence on their human foster parents and reintroduced to the jungle.

Cage beneath World Wildlife symbol.

Arrival at the rehabilitation center.

Part of the problem in preventing species from becoming endangered is that the people who are aware of such situations are more immediately pressed to solve the problems of other animals which are even closer to the brink of disaster. It would pay many a conservation program to learn how to safeguard some species before they reach internationally accepted danger levels.

Jane Goodall points out that of all the animals on earth it is to the chimpanzee in particular that we owe our greatest debt. We still have a great deal more to learn about ourselves and the study of chimpanzees has proved very rewarding so far. There is still a long way to go.

The chimpanzee reserve at Gombe Stream in Tanzania is now world famous and so too is the Budongo Forest in Uganda where much early behavioral research was carried out. But these areas form only a small part of the chimpanzee's natural range and those on the west coast of Africa have been less well researched. In 1978 Stella Brewer's *The Forest Dwellers* opened a new chapter for chimpanzees dwelling on that side of the African continent.

As had been done with orangs, Stella Brewer gradually worked herself into the position of setting up a full-time rehabilitation center for orphaned chimpanzees. Her first recruit, an emaciated juvenile close to death, was revived by an enthusiastic family in a house teeming with wildlife. The arrival of more animals soon brought matters to a head: they could not all

Educating Yula

'Bobo climbed the tree at that moment, came close to us and picked three pods. I praised him loudly for Yula's benefit. Bobo looked round at me in surprise and acknowledged the unusual praise by panting into my hair briefly on his way down the tree. Again I tried asking Yula, hoping Bobo's example might have made things easier for her to understand. Still I saw no gleam of comprehension in her eyes.

I picked up her hand and wrapped her fingers round a pod, my voice enthusiastic to indicate that she was doing the right thing. Then I held out my hand again, speaking to her all the time. She held the fruit briefly, then put her hand in mine, almost as if she were miming the action of giving me something. I put her hand round the pod again and placed her other hand on the small branch from which the pod hung. Then I placed my own hands over each of hers and, still talking encouragingly, tightened my grip around her hand on the pod and pulled it away from the branch. The short, tough stem by which the pod was attached tore away from the branch and Yula found herself holding a pod. I threw my arms around her and hugged her tightly in exaggerated praise. She was surprised at my excitement but obviously pleased. I held out my hand and Yula placed the pod in my palm. I hugged her again, food-grunting like mad. With Yula on my back I hurried down the tree, went immediately to a flat rock, hammered the pod open, and handed her the open pod full of flat, crunchy, purple seeds.

Yula food-grunted enthusiastically and ate. While she fed, I climbed the tree, broke a branch with six pods dangling from it, and brought this down to the ground. When Yula had finished, she hurried over to where I sat with the branch. I pointed to a pod and held out my hand. Yula briefly touched a pod and then mimed giving it to me. I shook my head, my voice slightly stern as I said: "No, give me the pod, Yula!" Again she touched the pod and looked away. I put her hand around the pod again, placed my hand over hers and pulled. The pod came away. I praised Yula again. I held my hand out, she gave me the pod and impatiently waited while I hammered it open for her as quickly as I could. For the third pod Yula needed almost no help. I pointed and held out my hand. Hesitantly she placed her hand on a pod and pulled it off by herself. I merely held the branch. I praised her profusely, hugging and kissing her, then, as soon as she held out the pod for me to take, I hammered it open for her and handed it back. We panted and food-grunted at each other and Yula climbed into my lap to eat her pod.

When she had finished I sat her down on the ground beside me. I picked a pod, walked the few yards to the tree and placed it in a fork. Yula was watching me. I walked back and sat down. Pointing at the pod in the fork, I asked her to get it for me. To my delight she got up almost immediately, went to the tree and came back food-grunting with the pod in her mouth. She handed it to me straight away and I opened it for her.

While she ate I took the branch with its one remaining pod to the tree and wedged it well into a fork. Yula was watching me. When she had finished eating I said nothing, hoping she would go and pick the last pod without any encouragement from me. She didn't but sat next to me rocking slightly. I knew if she didn't hurry Pooh and Bobo would take the pod so after a minute, I turned to Yula, pointed to the tree and held out my hand. She went over straight away, brought the whole branch with her and gave me the branch. For a second I wasn't sure what to do. I wanted her to pick the pod and wondered whether I would confuse her by insisting she did so after she had so confidently brought the branch to me. Finally I decided that it would confuse her more if I made exceptions and so I asked once more for the pod. She hesitated slightly, but after I'd pointed at the pod a couple of times, she pulled it off the branch.'

Stella Brewer, *The Forest Dwellers*

be kept at home indefinitely.

By chance a perfect area at the Abuko Nature Reserve, near the mouth of the Gambia River, presented itself in 1968 as the ideal place to keep all the animals. At that time there were just two chimpanzees, but more soon followed, for although Gambia has no wild chimpanzees itself, there is a market for those imported from neighboring countries, especially Senegal. The new recruits, from markets and dealers, came in varying conditions according to how long they had been captive and how well they had been treated. But again the day came when the developing chimps became too energetic to live under such conditions. Stella Brewer made one simple vow: that not one of them would go to a zoo. She wanted to release them all in the wild where she felt they rightly belonged. By degrees she persuaded the National Park authorities of Senegal to allow her to release her chimps in the Niokolo Koba Park. That was in 1972.

She soon realized how ill-equipped the apes were for self-survival and thus she found herself embarked upon the work which keeps her occupied today – that of teaching them how to be real chimpanzees, down to the recognition of the proper foodstuffs and learning which animals are dangerous to approach too closely. It worked; the chimpanzees became more independent of her and less afraid of every strange object they encountered on their own. Perhaps her greatest triumph was that of successfully introducing two young chimpanzees from London Zoo to the wilds of Senegal. In

Though chimpanzees are less threatened as a species than orangs and gorillas, overgrazing and overcultivation of potentially arid areas – like the one below – restricts their range.

1974 after a time with Jane Goodall at Gombe, she launched the Chimpanzee Rehabilitation Project as a full-time center for attempting to put back into the wild a small piece of what humans take out of it.

This brief survey does not encompass all the work that is being undertaken on the conservation of apes throughout the world today.

A world without apes would be a world poorer for humanity. Conversely, if the apes survive, their habitats will endure, the world will remain in possession of some of its primeval complexity and we shall have benefited not only our closest evolutionary relatives, but also ourselves.

The Gombe Stream National Park in Tanzania is – as a result of Jane Goodall's pioneering work – a sanctuary for chimpanzees.

A NURSERY FOR ORANGS

Since it was established in the early 1970s by two Swiss scientists, Regina Frey and Marika Borner, the Bohorok Rehabilitation Center in Sumatra has achieved its aims, and in a surprising way. It was originally designed to return to the jungle orphan orangs and pets (orangs were considered status symbols by officers and village chiefs). As these pictures of Regina Frey and her charges show, the foster parents built up an extraordinarily close relationship with their charges.

But the task proved both arduous and too limited in its aims. No more than a dozen or so orangs a year could be returned to the jungle and then with no certainty of long-term success. It was hard to identify forest which did not already have an established population of orangs, yet which contained enough food. It was uncertain whether the apes could breed. And what use was rehabilitation without tackling the root cause – unthinking deforestation and capture of young animals?

Consequently Bohorok has become a center more for the education of the Indonesians than for orang rehabilitation. Some 6000 tourists a year visit the center, and the publicity has ensured the steady dissemination of information about the threat both to orangs and to the forest. The orang, like the World Wildlife Fund's panda, has become a symbol of a wider issue. If the habitat is used correctly – it is argued – the orangs can be preserved in the best way possible, secure in the jungle of which they are a part. The center could well turn out to have a significance beyond the dreams of its founders.

Regina Frey feeds her orphaned orangs.

An ecstatic moment of reassurance.

A climbing lesson.

A quiet moment at play.

Acknowledgements

Cover Picturepoint
Back cover Survival Anglia Ltd/Lee Lyon
Title page Ardea London
6-7 Bruce Coleman Ltd/Lee Lyon
8 The Mansell Collection
10-11, 12, 13 Bruce Coleman Ltd
14-15 artwork Gary Hinks
16 Bodleian Library, Oxford
17, 18 (2 pictures) Royal Geographical Society
19 Fortean Picture Library
20 artwork Laurence Bradbury
20 Bruce Coleman Ltd/Norman Tomalin
21 *top* Dr Bernard Wood
21 *bottom* Dr Roger Fouts
22 The Mansell Collection
23, 24 (2 pictures), 25 (2 pictures) Dr Roger Fouts/Dave Rowe
26 Keystone Press Agency
27 NASA
28, 29 (3 pictures) Popperfoto
30-31 Brook Bond Oxo Ltd
32 (2 pictures), 33 *top left, top right, bottom left* Keystone Press Agency
33 *bottom right* Globe Photos Inc/Pierre Berger
34 Keystone Press Agency
35 Bruce Coleman Ltd
36 Anthro-Photo/D J Chivers
37, 38-9, 40 (3 pictures), 41 *bottom* Bruce Coleman Ltd
41 *top* Ardea London/Kenneth W Fink
42-3 Ardea London/C McDougal
44 (2 pictures), 46-7 Mary Evans Picture Library
48 (2 pictures) Reproduced by courtesy of the Trustees, The National Gallery, London
49 Anthro-Photo/David H Agee
50 Ardea London/C McDougal
51 Anthro-Photo/Irven DeVore
52, 53 Mary Evans Picture Library
55 Photo Researchers Inc
58-9 Anthro-Photo/David Agee
61 Anthro-Photo/Irven DeVore
63 Ilka Hanski
64, 65, 66 *left* Keystone Press Agency
66 *right* Globe Photos Inc
67, 68-9, 70 Photo Researchers Inc
71 Pictor International
72-3 Survival Anglia Ltd/Lee Lyon
74 Mary Evans Picture Library
76 Ardea London
76-7 Bruce Coleman Ltd/Lee Lyon
78 *top* The Mansell Collection
78 *bottom* Mary Evans Picture Library
79, 80-81 Historical Picture Service
81 Courtesy of Edgar Rice Burroughs Corp
82 Mary Evans Picture Library
83 *top* Globe Photos Inc
83 *bottom* Keystone Press Agency
84 Bruce Coleman Ltd/Lee Lyon
85 Photo Researchers Inc
88-9 Bruce Coleman Ltd/Lee Lyon
90 Globe Photos Inc
91 Keystone Press Agency
92 *top* Photo Researchers Inc
92 *bottom* Bruce Coleman Ltd/G D Plage
93 Zoological Society of London
95 Survival Anglia Ltd/Lee Lyon
96 Survival Anglia Ltd/Colin Willock
96-7, 98-9, 100, 101 (3 pictures) Survival Anglia Ltd/Lee Lyon
102-3 Anthro-Photo/Richard Wrangham
104 The Mansell Collection
105 Mary Evans Picture Library
106-7 Photo Researchers Inc
108-9 (3 pictures) Anthro-Photo/Richard Wrangham
110 Ardea London/Kenneth W Fink
110-11 Photo Researchers Inc
113 Keystone Press Agency
114-15 Anthro-Photo/Richard Wrangham
116, 117 Anthro-Photo/Nancy Nicolson
118 Anthro-Photo/Richard Wrangham
119 Mary Evans Picture Library
120 Picture Researchers Inc
122-3 Anthro-Photo/Irven DeVore
124, 126-7 Anthro-Photo/Richard Wrangham
128-9 (2 pictures) Anthro-Photo/James Moore
130, 131, 132, 133, 134, 135, 136, 137 Anthro-Photo/Richard Wrangham
138-9 Mary Evans Picture Library
140 Anthro-Photo/Irven DeVore
140-41 Anthro-Photo/Richard Wrangham
143, 145 Popperfoto
146-7 Dr Bernard Wood
148 Mary Evans Picture Library
150 *top* The Mansell Collection
150 *bottom* Radio Times Hulton Picture Library
151 (2 Pictures) Mary Evans Picture Library
153 *bottom* Radio Times Hulton Picture Library
154, 155 *bottom* Popperfoto
155 *top The Illustrated London News*
156, 157 (6 pictures), 158 Dr Bernard Wood
159 (2 pictures), 160 Radio Times Hulton Picture Library
161 Historical Picture Service
162-3 (4 pictures) Dr Bernard Wood
164-5 (3 pictures) *The Illustrated London News*
166 The Australian Information Service, London
167 Mary Evans Picture Library
168-9 (2 pictures) *The Illustrated London News*
170-71, 172, 176, 177 Bruce Coleman Ltd
180 Ardea London/M E Gore
181 (3 pictures) Bruce Coleman Ltd
183 Ardea London/M E Gore
184 Anthro-Photo/Irven DeVore
185, 186-7, 188-9 (3 pictures) Bruce Coleman Ltd
All uncredited pictures are courtesy of John Man Books
Index & picture research – Susan de la Plain.

Index